Starch
Structure and Functionality

Starch
Structure and Functionality

Edited by

P.J. Frazier
Dalgety PLC, Food Technology Centre, Cambridge, UK

P. Richmond
School of Information Systems, University of East Anglia, Norwich, UK

A.M. Donald
Cavendish Laboratory, University of Cambridge, UK

THE ROYAL
SOCIETY OF
CHEMISTRY
Information
Services

The proceedings of an International Conference sponsored by the Food Chemistry Group of The Royal Society of Chemistry in association with the Institute of Food Science and Technology Research Subject Group held at the University of Cambridge, UK, 15–17 April 1996.

Special Publication No. 205

ISBN 0-85404-742-5

A catalogue record for this book in available from the British Library

Published by The Royal Society of Chemistry,
Thomas Graham House, Science Park, Milton Road,
Cambridge CB4 4WF, UK

Printed by Bookcraft (Bath) Ltd

Preface

This book presents the proceedings of the meeting Starch: Structure and Function which was held at Churchill College, Cambridge from 15-17 April 1996. The meeting followed in the footsteps of two other recent meetings which covered different aspects of current developments in our understanding of starch, i.e. "Wheat Structure, Biochemistry and Functionality" held in Reading from 10-12 April 1995 and "Chemistry and Physics of Baking: Materials, Processes and Products", held in Sutton Bonington (Nottingham University) from 10-12 April 1985 (RSC Special Publication No. 56, 1986). This particular meeting was arguably unique in that physicists, chemists, biologists and medical experts rubbed shoulders throughout the whole meeting and the quality of the talks was high from a scientific standpoint. However care was also taken to ensure that they were understandable by scientists from outside the immediate discipline - a feat usually rarely achieved.

Starch is the basis of a huge global industry. Over two thirds of the industry is directed towards exploitation of starch for non-foods. Even so, the remainder represents billions of tonnes per annum of material from a range of agricultural products which finds its way into an increasing range of prepared food products. It was to the food side that this conference, sponsored jointly by the Royal Society of Chemistry and the Institute of Food Science, directed its programme. Keynote speakers focused the attention of delegates onto three key areas:

- diet and health,
- physico-chemical aspects of starch,
- genetic and agronomic questions.

The keynote speakers were asked to provide a fairly broad overview and, where possible to pitch their talk at a level which could be understood by all the conference attendees. The reader will notice in this book that these keynote contributions are different in style to the more focused research contributions. In addition we have included abstracts of all the poster sessions. The editors sincerely hope that this all adds to the total value of the book.

Both the venue, Churchill College, and the weather proved to be first class. The meals, and especially the conference dinner, were excellent with thanks going to Dalgety who helped subsidise the hospitality. Dalgety Chief Executive, Richard Clothier, gave an excellent after dinner presentation to accompany the wine. The Biotechnology & Biology Research Council also provided additional support for the meeting which was attended by over 150 delegates from all over the globe. Indeed the meeting was judged to be so successful that plans for a follow up meeting to be held on 27-29 March 2000 at the same venue are already under way.

Peter Richmond, University of East Anglia, Norwich, NR4 7TJ
Athene Donald, Cavendish Laboratory, University of Cambridge, CB3 OHE
Peter Frazier, Dalgety Food Technology Centre, Cambridge, CB1 2JN.

Contents

POSTER PRESENTATIONS

STRUCTURE/FUNCTION RELATIONSHIP OF STARCHES IN FOOD

P. J. Lillford and A. Morrison

Unilever Research
Colworth Laboratory
Sharnbrook, Bedford, MK44 1LQ

1 INTRODUCTION - WHY STARCH?

As the storage energy source of many seeds, roots and tubers it is cheap and ubiquitous. It provides soluble macromolecules exhibiting properties typical of such materials. i.e. high viscosity, adhesion, surface coating.

These properties are highly desirable throughout industry but starch has another advantage. As a polymer of α-glucose it is readily digested and forms the major energy source in most diets. 80% of all food crops comprise cereal and starchy roots.

Its value as a material throughout industry has led to its production as an ingredient. Annual production figures are difficult to obtain but Table 1 shows the scale of the operations.

Table 1 *Production and Consumption*

Production

World Production 1992/93	27.5 million tonnes
EEC Production 1992/93	6.6 million tonnes
Japan 1990	2.5 million tonnes

Consumption   1993

EEC		U.S.A.
Food	31%	30%
Industrial	69%	70%
		3×10^6 tonnes

In its native granular form, starch has few uses. To release the polymer properties, granule disruption and sometimes also modification are necessary. These can be achieved by chemical and/or physical processes. Many combinations have been tried so that starch as an industrial ingredient comes in many forms. (Tables 2 & 3)

It would be simple to describe why such a variety exists had the development been systematic and scientifically based. However, this is not the case and successful ingredients are a result of systematic empiricism.

Table 2 *Starch and Starch Derivatives in Commercial Use*

Native Starches	CORN/MAIZE
	WHEAT
	SORGHUM/MILO
	RICE
	POTATO
	TAPIOCA/CASSAVA
	SAGO
	ARROWROOT
	PEA
Genetically modified cultivars	WAXY MAIZE
	AMYLOMAIZE
	WAXY SORGHUM
	WAXY RICE
Chemically modified esters, ethers, cross-linked, cationic, oxidised	MONOPHOSPHATES
	ACETATES
	HYDROXYPROPYL
	ADIPATES
	DIPHOSPHATES
Hydrolysed starches - acid, enzyme	CORN SYRUPS
	MALTODEXTRINS
	GLUCOSE
Dextrins - low moisture, heat, acid treatments	YELLOW GUMS
	WHITE GUMS
	BRITISH GUMS

2 STARCH USAGE IN FOODS

It is not possible in this lecture to provide an analytical explanation of how all these forms of starch function. We will try, at least, to put some "after the fact" explanation into the use of starches as <u>food</u> ingredients. Even then a wide variety of products is available, each tailored for specific applications.

3 THE PROPERTIES OF NATIVE STARCH

3.1 Let us begin with the botanical form

Starch is deposited in granular storage bodies in the plastids of most higher plants. The size and size distribution varies with botanical origin, but all are comprised of predominantly two polysaccharides, amylose and amylopectin and remarkably little of anything else. Note that the precise nature of both macromolecules varies from source to source e.g. degree of branching, phosphate substitution. Table 4 shows the composition of typical wheat starch granules. The function of the minor components are poorly understood but are involved in maintaining structural integrity of the granule and can therefore significantly influence its disruption in use.

Tables 3 *Uses of Starch and Derivatives*

<u>Food</u> <u>Uses</u>		<u>Industrial</u> <u>Uses</u>	
	sauces		paper and board
	soups		textiles
	dressings		plastics
	baked goods		rubber
	dairy products		oil
	meat products		pharmaceuticals
	drinks		medicine
	ice cream		cosmetics
			adhesives
	refrigerated		sewage and water treatment
	deep-frozen		alcohol
	dry mix		
	tropical storage - liquids (30°)		
	ambient (temperate)		sizing
			coating
	thickening		texturising
	gelling		viscosity control
	stabilising		flocculation
	sweetening		ion exchange
	bulking		adhesive
	texturising		dusting
	fat replacement		fuel

Table 4 *Composition of Wheat Starch Granules*

Amylose			23 - 27%
Amylopectin			73 - 77%
Lipids (wheat)	-	on surface	0.02 - 0.6%
	-	in interior	0.1 - 0.2%
Proteins (wheat)	-	on surface	0.006 - 0.5%
	-	in interior	0.07%

The granule itself has a highly organised architecture of concentric rings which correspond to the dimensions of the constituent polysaccharides so that composition and structure appear to be closely related.

Under polarised light, the granules exhibit birefringence from which long range crystalline order within and between polymer chains can be deduced.

3.2 Molecular structure and hydration of granules

The structure of the component polysaccharides has been extensively studied but because of their hydrophilic nature, the conformation of soluble polymers is different from that in the unhydrated granule. In their botanical, natural forms, the consensus is that amylose is present as both amorphous material and as a hollow helical tube, with lipid complexed internally i.e. its secondary structure is of a linear polymer with few branches, of molecular weight ~10^6 Daltons.

The consensus structure for amylopectin in a dendritic highly branched polymer, of molecular weight up to 4 x 10^8D making it one of the largest naturally occurring molecules.

To obtain the benefits of the polymer functionality it is necessary to extract the polymers with minimal damage. This is done with a compatible solvent which for industrial uses can be any hydrophilic solvent. In food use this is always water for both cost and safety reasons.

Granule disruption has been widely studied by various techniques. Cooperative swelling occurs as a function of temperature, but different events are detected depending on the observation technique. (Table 5)

Table 5   *Starch Structure and Gelatinisation*

STARCH	CRYSTALLINE[a] ORDER (%)	MOLECULAR[b] ORDER	TEMPERATURE OF BIREFRINGENCE[c] LOSS (C)	ONSET OF[d] MELTING (°C)
WHEAT	20	39	52 - 61	54
MAIZE	27	43	62 - 70	68
POTATO	24	40	56 - 67	58
WAXY MAIZE	28	48	62 - 70	62
TAPIOCA	24	44	58 - 70	66
AMYLOMAIZE			95 - 130	

a   Determined by x-ray diffraction ($\pm$ 2%)
b   Determined by ^{13}C NMR (cp-mas) ($\pm$ 2%)
c   Determined using hot-stage (Kofler) microscope
d   DSC

It is clear that order is lost at the level of Bragg diffraction spacings at lower temperatures than molecular mobility or the overall granule swelling associated with heat absorption.   Thus the process either has a low degree of cooperativity or is microheterogeneous within the granule. All of these processes take place before total solubility of the granule polymers. This is not surprising when molecular sizes and the inherent structural complexity of the native granule is considered. The self diffusion coefficient of molecules of these shapes and sizes is extremely low so that in most processes occurring in reasonable time scales (mins to hours), mechanical stirring is necessary and it is here where the problems start.   (Table 6)

3.3 Mechanical disruption of granules

In normal practice where simultaneous shear and temperatures is used as the disruption process, viscosity rises to a maximum and subsequently falls.

The common explanation for this is that maximum viscosity is reached at maximum granule swelling. Thereafter viscosity falls as granules fragment and polymers dissolve.

However, this cannot be the whole story, and a comprehensive review of the structural changes occurring during heat and shear processing of starches has been presented recently. (Hermansson et al)

Table 6 *Effect of shear on different Starches, - Resultant viscosity (mPas) measured at 100s^{-1} (Haake Rotovisco)*

Starch (5%)	Light Shear	Heavy Shear
Tapioca	136	24
Rice	271	34
Wheat	150	34
Maize	160	53
Waxy Maize	356	32
Snowflake 06307 (wm)	262	155
Thermflo (wm)	97	149
Pureflo (wm)	70	146
Purity D 'A' (tapioca)	39	53
Purity 69A (tapioca)	24	44
Thermtex (wm)	10	155

In commercial practice, where simultaneous heat and shear is usually performed, an alternative proposition is that viscosity increases until maximum swelling and hydration occurs, further shear not only solubilises but also shears the amylopectin molecules. The large drop in molecular weight of amylopectin is thus the origin of the subsequent viscosity drop.

Evidence for the latter is given from granule disruption experiments where shear is kept to a minimum.

Under these conditions, granule disruption is extensive but no viscosity drop is observed even up to 130°. In fact, maximum viscosities are reached after heating to circa 100°. The structures of these "solutions" are, in fact, phase separated mixtures of amylose and amylopectin. Under low or no shear conditions, the system is probably bicontinuous in amylose and amylopectin, whereas after high shear the amylopectin becomes continuous, with inclusion of almost spherical amylose inclusions.

I.e the drive towards phase separation is assisted by the reduction in viscosity of amylopectin by shear degradation of its molecular size.

Further evidence for the shear induced damage of amylopectin is provided by studies of the refined polymer and a waxy maize starch (predominately amylopectin). (Table 7)

Table 7 *Effect of shear, during prepared and measurement, on the intrinsic viscosity [η] of waxy maize starch and amylopectin extracted from waxy maize starch*

Sample	Intrinsic Viscosity [η] ml g^{-1} Measured at Shear Rate Indicated		
	1 sec^{-1}	100 sec^{-1}	600 sec^{-1}
Extracted amylopectin (0.5%)	109	-	-
Extracted amylopectin (5.0%)	67	8	1
Waxy maize starch - minimal shear hydration	107	23	5
Waxy maize starch - high shear hydration	38	1	0.5

In each case the intrinsic viscosity of virtually unsheared systems is at least twice as high as moderately sheared systems. This implies a molecular weight reduction of an order of magnitude even at these shear rates. Banks et al measured comparable molecular weight reductions when amylopectin is sheared at concentrations greater than 0.5 wt%.

Thus the effect of shear on starches containing both amylose and amylopectin is two-fold. Firstly, the discontinuous amylose phase is concentrated into small, more uniform particles, but more importantly, the molecular weight of the continuous amylopectin is dramatically reduced.

It is ironic that because of the necessary speed of commercial or kitchen processing, the true viscosifying capability of amylopectin is rarely achieved.

4 THE NEED FOR MODIFIED STARCHES

Not all products are made or consumed immediately after starch gelatinisation. Subsequent cooling, storage in chill or even freezing is common. Under these conditions, the gels of native starch exhibit shrinkage, water syneresis and as a result the texture can become tough. The effect is exacerbated by further heating. This phenomenon of "set-back" otherwise known as retrogradation, is due to the crystallisation of amylose, and some of the amylopectin. The time course for the latter molecule is much slower, so that waxy starches are preferred in product to be re-heated before consumption.

We have mentioned previously the shear damage to which amylopectin is so sensitive, and this inhibits the use of native waxy maize starch as an ingredient in many applications. This can be countered by chemical crosslinking of the glucan chains, normally using chemical agents such as phosphorus oxychloride, sodium trimetaphosphate and adipic anhydride. Figure 1 shows the stabilising effect of low level of crosslinking, but at high levels the swelling rate is reduced and the intrinsic viscosity of the native unsheared amylopectin is never achieved.

Figure 1.

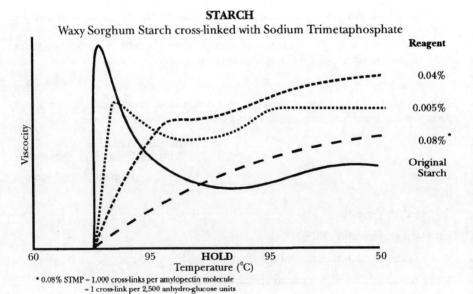

STARCH

Waxy Sorghum Starch cross-linked with Sodium Trimetaphosphate

* 0.08% STMP = 1,000 cross-links per amylopectin molecule
= 1 cross-link per 2,500 anhydro-glucose units

Unfortunately, whilst shear sensitivity is reduced by crosslinking, the resultant starch can become more sensitive to freeze damage and syneresis. The introduction of hydrophilic substituents (acetate, hydroxy propyl and phosphate) limit the recrystallisation (retrogradation) by steric interference so that acetylated distarch adipate retains viscosity and is cooling stable. However, chemical modification renders the ingredient far from "natural".

In summary, the properties of modified starches are described in Figure 2.

Figure 2

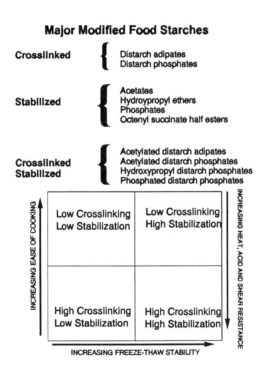

5 IN SUMMARY

Starch should be considered a material rather than a simple ingredient. By understanding the enormous variety in structure and composition of natural sources its properties can be optimised in use. Even then, the propensity of the constituent polymers to phase separate, rupture under shear and retrograde limits their use.

We have seen that the uses of starches without chemical modification are greatly restricted because of retrogradation of amylose and shear damage to amylopectin, but the use of chemically modified starches is unpopular with sophisticated consumers and restricted by legislation.

How can we proceed to gain the greatest use of starches in foodstuffs? A number of ways forward can be considered:

- There are many alternative sources of starch which have hardly been examined and which may yield interesting properties e.g. starches from leguminous plants.
- Use of genetically modified starches, of which waxy maize itself is an example. High amylose starches are available which give greater gelling properties to starch.
- Use of phase separation with other macromolecules e.g. gelatin, carrageenan. There are examples in the literature of phase separation between macromolecules and materials formed from starch by heat, acid and enzyme treatment e.g. Paselli SA2. Products in which gelatin occupies the discontinuous phase have the melting characteristics of gelatin but not the some three dimensional gel network. Gel strength of κ-carrageenan or locust bean gum and carrageenan gels can also be controlled by starch levels.
- Recent work has suggested that the protein molecules on the surface of granules plays an important role in controlling the swelling of granules, which can be modified or enhanced by protein manipulation.
- Lipids have long been used to control retrogradation by forming complexes with amylose as noted earlier. Further combinations could be considered.

Despite these advances there seems no way at present to ensure the stability of hydrated amylopectin molecules against mechanical degradation without chemical modification. Process modification to minimise shear disruption must also be considered.

6 BIBLIOGRAPHY

1. "Chemistry and Industry of Starch", R.W. Kerr, Ed., Academic Press, N.Y., 1950.
2. "Methods in Carbohydrate Chemistry", Vol. IV; in "Starch", R.L. Whistler, Ed., Academic Press, N.Y., 1964.
3. "Starch and Its Components", Banks, W. and Greenwood, C.T., Halsted Press, N.Y. 1975.
4. "Industrial Uses of Starch and its Derivatives", J.A. Radley, Ed., Applied Science Publishers, London, 1976.
5. "Starch: Chemistry and Technology", R.L. Whistler, J.N. Miller and E.F. Paschall, Ed., Academic Press Inc. London, 1984.
6. "Starch, a Phase Separated Biopolymer System", A-M., Hermansson, S. Kidman, and K. Svegmark, in "Biopolymer Mixtures", S.E. Harding, S.E. Hill and J.R. Mitchell, Ed. Nottingham University Press, 1995.

STARCH AND HEALTH

H. N. Englyst and G. J. Hudson

MRC Dunn Clinical Nutrition Centre
Hills Road
Cambridge CB2 2DH

1 INTRODUCTION

Starchy foods are the world's most abundant staples, and the site and rate of digestion of starch, which are highly dependent on processing, have significant implications for health. The human digestive system is evolutionarily adapted to cope with the diet that prevailed until the Agricultural Revolution about 10,000 years ago.[1] In comparison with the modern "Western" diet, the pre-agriculture diet contained less fat and far more dietary fibre (plant cell walls). Starch in the pre-agriculture diet was derived mainly from roots, beans, fruits and tubers; cereal grains were not a major component of this diet. The natural encapsulation of starch within undamaged plant cell walls (dietary fibre) in the raw or only lightly processed foods typical of that diet slows the rate of digestion, resulting in a sustained release of glucose. Modern starchy food products are mainly cereal-based and often finely milled, which disrupts the plant cell walls, and the starch is often fully gelatinised during processing. The release of starch from within the cell walls, which may be removed during refining, and the gelatinisation of the starch leads to rapid digestion and absorption of the starch in the small intestine, contrary to the fate of the starch in the pre-agriculture diet. There are strong indications that the large amounts of rapidly available glucose derived from starch and free sugars in the modern diet, in combination with the consumption of discrete meals, lead to periodic elevated levels of plasma glucose and insulin that are detrimental to health in many contexts, including diabetes, coronary heart disease, cancer and ageing.[2-5] The evolutionary time-scale is too long to expect that there has been any significant adaptation of the human gut to the cereal-based diets introduced during the Agricultural Revolution and certainly not to the highly processed foods in the modern diet.[6] Current knowledge of the fate of dietary starch means that the potentially undesirable properties of many modern starchy foods could be altered by using processing techniques that yield foods with a reduced rate of starch digestion in the small intestine. Such products would more closely resemble the foods in the pre-agriculture diet with respect to the rate of digestion of the starch that they contain, which would be of great benefit to public health.

This paper examines the main differences between the starch in the pre-agriculture diet and modern starchy foods. It points out the detrimental effects to health of rapidly available carbohydrate, and possibly of increased amounts of starch not digested in the small intestine. It proposes a nutritional classification of dietary carbohydrates and makes suggestions as to how the likely glycaemic response to foods can be monitored and the potential benefits that would follow if modern starchy foods can be made more compatible with health.

Table 1 *Summary of Dietary Carbohydrate Classification*

Class	Components	Comments
Sugars		
	Mono- and disaccharides	Links with diabetes and coronary heart disease, and with the ageing process *via* protein glycation
	Sugar alcohols	Sparingly absorbed. Partly metabolised
Short-chain carbohydrates		
	Oligosaccharides Inulin	May be fermented in the large bowel. Inulin and fructo-oligosaccharides have been shown to stimulate growth of bifidobacteria
Starch		
	Rapidly digestible starch (RDS) (includes maltodextrins)	RDS and RAG* associated with elevated plasma glucose and insulin. Links with diabetes and coronary heart disease, and with the ageing process
	Slowly digestible starch (SDS)	Moderate influence on plasma glucose and insulin levels. Nutritionally the most desirable form of starch
	Resistant starch (RS)	Desirability of increase in foods requires further evaluation
Non-starch polysaccharides (NSP)		
	Cell-wall NSP in unrefined plant foods (dietary fibre)	Encapsulate and moderate rate of digestion and absorption of sugars and starch. Marker for the high-fibre diet rich in fruit, vegetables and whole-grain cereal products recommended in the national guidelines
	Other NSP	No encapsulation of other nutrients. Not a marker for the high-fibre diet recommended in the national guidelines Not dietary fibre

* Rapidly Available Glucose (RAG) = RDS + free glucose + glucose from sucrose.

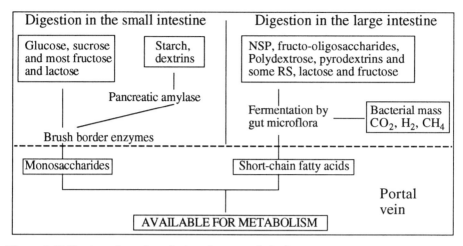

Figure 1 *Utilisation of starch and other dietary carbohydrates*

2 CLASSIFICATION OF DIETARY CARBOHYDRATES

The human diet contains a wide range of carbohydrates and it is essential to have a chemically based classification and specific analytical methodology if useful values are to be obtained. Food carbohydrates are classified into groups according to: degree of polymerization (DP); identity of constituent sugars; and type(s) of glycosidic linkages present. Table 1 shows the classification of food carbohydrates, their likely fate in the small intestine and the magnitude of the glycaemic response. Classification of carbohydrates on the basis of chemically identified components provides also a logical basis when considering the utilisation of food carbohydrates (Figure 1).

Groups may be combined or subdivided and, based on a series of studies in man,[7-9] we propose a classification of dietary starch that reflects the different rates and extents of digestion, and absorption of the products, in the small intestine (Table 2). Rapidly Digestible Starch (RDS) is starch that is rapidly and completely digested in the small intestine; Slowly Digestible Starch (SDS) is starch that is completely but more slowly digested in the small intestine; and resistant starch (RS) is defined as the sum of starch and starch degradation products that, on average, reach the human large intestine.[10-13] Most starchy foods contain starch belonging to two, if not all three, categories of starch, but in very different proportions. The three principal causes of incomplete starch digestion are reflected in the three subfractions of RS (Table 2).

The three main types of starch and the three subfractions of resistant starch can be measured *in vitro* (see later).

Rapidly Available Glucose (RAG) is the sum of free glucose, glucose from sucrose and glucose from RDS. RAG values are obtained by the methodology outlined in Figure 2 and are expressed as g/100g of food as eaten. These values may be used for direct comparison of foods and for calculation of the total RAG values of meals and diets. The RAG values represent the amount of glucose likely to be rapidly absorbed in the human small intestine.[14]

3 FACTORS AFFECTING THE RATE AND EXTENT OF STARCH DIGESTION AND ABSORPTION

All dietary starch is potentially degradable by the action of alpha-amylase. However, certain factors can reduce the rate at which starch is hydrolysed and absorbed *in vivo*,[11] thus delaying the appearance of glucose in blood after a meal. For some foods, hydrolysis by

Table 2 *Nutritional Classification of Starch*

Class	Example of Occurrence	Site of Digestion and Absorption	Glycaemic Response
Rapidly digestible starch	Processed foods	Small intestine	Large
Slowly digestible starch	Muesli, legumes, pasta	Small intestine	Small
Resistant starch			
Physically inaccessible	Whole grains		
Resistant granules	Unripe banana	Large intestine (fermented)	None
Retrograded amylose	Processed foods		

amylase is hindered to such an extent that some starch passes into the colon, which may result from one or more of the following principal causes. Starch may be physically entrapped (RS1), which hinders contact with amylase, e.g. starch within a whole cereal grain or in pasta. Raw starch may occur in a granular form in which the crystal structure is particularly resistant to amylase (RS2), e.g. starch granules in banana and potato. Cooked starch, once cooled, may contain regions where the starch chains (mainly amylose) have recrystallised or retrograded into a configuration that is highly resistant to pancreatic amylase (RS3). These factors are related to the starchy food itself and may be described as intrinsic factors. In addition to these intrinsic factors, starch digestion is influenced by physiological, extrinsic, factors. Extrinsic influences include the degree of chewing, concentration of amylase in the gut and transit time through the stomach and small intestine, and for these reasons the rate and extent of starch digestion *in vivo* is variable both within and between individuals. The following measurements therefore aim to represent the population average.

4 MEASUREMENT OF RAG AND STARCH

The procedure used to measure rapidly digestible, slowly digestible and resistant starch fractions has been described in detail elsewhere.[11] In short, the various categories of starch and rapidly available glucose (RAG) are measured after incubation with a mixture of invertase (to hydrolyse sucrose), pancreatic alpha-amylase and amyloglucosidase at 37°C. The incubation is done in capped tubes immersed horizontally in a shaking water-bath with a stroke length and speed calibrated to yield predetermined values for a reference material (raw potato starch). All foods that normally require chewing are minced by a standard procedure. The incubation tubes contain balls, which disrupt the food particles, and guar gum is added to standardise the viscosity of the incubation mixture. Foods that are normally eaten warm are cooked immediately before they are taken into the analysis. A value for RAG is obtained as the glucose released from the food after 20 minutes (G_{20}). A second measurement (G_{120}) is obtained as the glucose released after a further 100 minutes incubation (120 minutes incubation *in toto*). A third measurement (total glucose) is obtained by gelatinisation of the starch in boiling water and treatment with 2 M potassium hydroxide at 0°C to disperse any retrograded starch, followed by complete enzymic hydrolysis with amyloglucosidase. Resistant starch is measured *in vitro* as the starch that remains unhydrolysed after a total of 120 minutes incubation. Separate values can be obtained for different types of resistant starch; physically inaccessible starch (RS1), resistant starch granules (RS2) and retrograded starch (RS3). Studies with ileostomy patients have shown good agreement between the amount of resistant starch measured *in vitro* and the average amount of starch that is not digested in the human small intestine.[11, 13, 15]

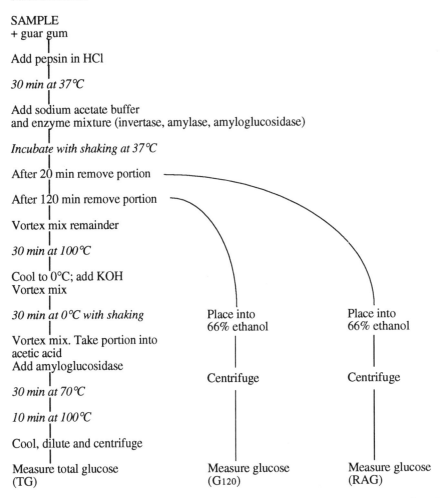

SAMPLE
+ guar gum

Add pepsin in HCl

30 min at 37°C

Add sodium acetate buffer
and enzyme mixture (invertase, amylase, amyloglucosidase)

Incubate with shaking at 37°C

After 20 min remove portion

After 120 min remove portion

Vortex mix remainder

30 min at 100°C

Cool to 0°C; add KOH
Vortex mix

30 min at 0°C with shaking   Place into   Place into
                                      66% ethanol   66% ethanol
Vortex mix. Take portion into
acetic acid
Add amyloglucosidase   Centrifuge   Centrifuge

30 min at 70°C

10 min at 100°C

Cool, dilute and centrifuge

Measure total glucose   Measure glucose   Measure glucose
(TG)   (G120)   (RAG)

Figure 2 *A summary of the analytical strategy for measurement of rapidly available glucose (RAG), the glucose released by 120 minutes of incubation (G120), and total glucose (TG) in the same sample. See the text for details.*

Figure 2 shows the analytical scheme for the measurement of total glucose (TG; the sum of free glucose, the glucose part of sucrose and glucose derived from starch and maltodextrins), rapidly available glucose (RAG) and the glucose released after 120 minutes of incubation (G120). A value for free glucose (FG), including the glucose released from sucrose, may be obtained separately. The calculations for rapidly digestible starch (RDS), slowly digestible starch (SDS) and Total starch (TS) are as follows:

RDS = 0.9 x (RAG – FG);   TS = 0.9 x (TG – FG);   SDS = 0.9 x (G120 – RAG)

where 0.9 is the factor to convert monosaccharide (glucose) to polysaccharide (starch). Resistant starch (RS) values may be calculated without the need to measure free glucose:

RS = 0.9 x (TG – G120)

The RAG measurement takes only two to three hours, and RS can be measured well within a working day. With the distinction between RDS and SDS defined by the *in vitro* methodology, starch fractions and the RAG values were determined for a range of starchy foods (Table 3).

Table 3 *Analysis of 39 Starchy Foods* (TG, FG and RAG, g monosaccharide/100g of food as eaten; TS, RDS, SDS and RS, g starch/100g of food as eaten; Sol, Insol and Tot NSP, g polysaccharide/100g of food as eaten)

	DM	TG	FG	TS	RDS	SDS	RS	RAG	NSP Sol	NSP Ins	NSP Tot
Cereals											
Buckwheat	30.3	24.7	0.1	22.1	11.8	8.5	1.8	13.2	0.4	0.4	0.8
Pearled barley	23.3	19.1	0.1	17.1	8.0	7.0	2.1	9.0	1.6	3.2	4.8
Sweet corn	28.8	19.7	0.7	17.1	15.4	1.4	0.3	17.8	0.6	2.7	3.3
Bread											
Rye wholemeal	63.1	38.9	1.4	33.8	23.2	7.4	3.2	27.2	4.2	4.2	8.4
Ryvita crispbread	93.8	67.4	1.1	59.8	48.8	6.7	4.3	55.3	3.8	7.7	11.5
Wheat white	61.2	46.4	0.1	41.7	37.4	3.7	0.6	41.7	1.0	0.7	1.7
Wheat wholemeal	58.1	39.2	0.4	35.0	32.1	1.4	1.5	36.1	1.8	3.5	5.3
Biscuits											
Digestive	98.0	59.8	8.1	46.5	32.0	12.6	1.9	43.7	1.1	1.1	2.2
Oatmeal	98.6	62.6	0.4	55.9	48.8	6.2	0.9	54.6	4.1	3.4	7.5
Rich tea	98.4	62.9	8.7	48.8	38.6	8.9	1.3	51.6	1.1	0.6	1.7
Water	95.5	78.0	0.4	69.8	65.4	3.8	0.6	73.1	1.7	1.2	3.0
Breakfast cereals											
All Bran	86.0	37.2	12.4	22.2	20.6	0.5	1.1	35.3	3.6	18.1	21.1
Oat bran	95.6	51.8	1.0	45.8	31.2	13.6	1.0	35.7	8.0	5.1	13.1
Porridge Oats	23.6	14.6	0.1	13.0	9.9	3.1	0.1	11.1	0.9	0.7	1.7
Puffed Wheat	90.8	77.4	1.0	68.7	62.5	0.0	6.2	70.4	2.3	5.3	7.5
Rice Krispies	96.5	85.0	7.4	69.8	65.6	1.7	2.5	80.3	0.1	0.4	0.5
Shredded Wheat	91.7	70.0	0.9	62.2	48.7	11.9	1.6	55.0	2.0	7.8	9.8
Weetabix	94.2	65.5	2.2	57.0	56.8	1.0	0.0	65.3	3.1	6.6	9.7
Cooked rice											
Brown – long grain	37.7	26.4	0.1	23.8	14.6	9.2	0.0	16.3	0.0	0.8	0.8
White – long grain	32.5	25.6	0.0	23.0	17.4	5.6	0.0	19.3	0.0	0.2	0.2
White – parboiled	31.2	31.0	0.1	27.8	16.6	10.0	1.2	18.5	0.0	0.2	0.2
Pasta											
Macaroni	31.0	29.3	0.2	26.2	13.4	12.0	0.8	15.1	0.5	0.4	0.9
White spaghetti	31.7	26.2	0.1	23.5	13.5	9.0	1.0	15.1	0.5	0.5	1.0
Legumes											
Beans in tomato sauce	24.0	9.6	0.5	8.2	5.5	1.2	1.5	6.6	2.0	1.2	3.2
Butter beans	33.0	13.0	0.3	11.4	9.4	0.8	1.2	10.7	2.4	3.4	5.9
Chickpea	33.5	18.4	0.2	16.4	5.1	8.8	2.5	5.9	1.4	2.8	4.2
Chickpea (canned)	31.0	17.8	0.3	15.8	9.2	3.8	2.8	10.5	1.3	2.6	3.9
Haricot beans	32.6	20.5	0.3	18.2	4.1	5.8	8.3	4.9	3.0	3.7	6.6
Kidney beans	33.0	19.3	0.4	17.0	4.7	9.8	2.5	5.6	3.0	3.3	6.3
Kidney beans (canned)	32.0	16.9	0.7	14.6	7.6	5.2	1.8	9.1	2.9	3.2	6.1
Lentils – red	27.4	17.8	0.2	15.8	7.3	6.1	2.4	8.3	0.5	1.1	1.6
Peas – frozen	21.8	9.2	1.2	7.2	4.1	1.0	2.1	5.8	1.6	3.6	5.2
Peas – marrowfat	30.7	17.5	0.4	15.4	7.4	5.0	3.0	8.6	1.4	3.3	4.6
Pinto beans	40.1	18.3	0.4	16.1	9.3	5.0	1.8	10.7	3.2	3.6	6.8
Other											
Potato – boiled	21.3	18.0	0.2	16.0	15.2	0.7	0.1	17.1	0.7	0.7	1.4
Potato – crisps	98.2	56.0	0.4	50.0	42.7	2.8	4.5	47.8	1.0	1.2	2.1
Potato – instant	24.2	14.4	0.3	12.7	10.9	1.1	0.8	12.4	0.8	0.8	1.6
Sweet potato	24.3	12.6	2.3	9.3	7.5	0.8	1.1	10.6	0.9	1.1	2.0
Yam	32.6	20.3	1.7	16.8	14.3	0.4	2.1	17.6	0.6	0.7	1.2

5 ANALYSIS OF STARCHY FOODS

Table 3 shows the dry matter (DM), TG, FG and TS content, the proportions of the various fractions of starch, the RAG values and the NSP content for a range of foods. (The dry matter values for the breakfast cereals do not include any milk or sugar that may be added, since this is highly variable.) The legumes have the lowest RAG values, which are associated with low levels of free sugar and a low proportion of starch measuring as RDS. The cause of the slow and incomplete digestion of legume starch is probably a combination of starch granules being encapsulated by dietary fibre (plant cell walls) and not being fully gelatinised. Spaghetti, macaroni and pearled barley are examples of foods with moderate RAG values as the result of a dense structure and a low dry matter content. For digestive biscuits, a considerable proportion of the starch is measured as SDS due to incomplete gelatinisation but the high dry matter content results in a higher RAG value for digestive biscuits than for white bread. Cooked potato has a low RAG value because of the low dry matter content. The RAG value of 80 for Rice Krispies is markedly higher than those of the other breakfast cereals shown in Table 3.

6 COMPARISON OF RAG AND GI

The starch and the NSP (dietary fibre[16]) values for the 39 starchy foods were expressed on the same basis (per 100g of "available carbohydrate") as the glycaemic index (GI) and compared by correlation analysis with GI values taken from the literature for 23 cereal foods and 16 other types of food.[17,18] As expected, when all the foods are considered together (Table 4), there is no significant (NS) correlation between GI and total carbohydrate, a negative correlation between GI and NSP, slowly digestible starch and resistant starch, and a significant positive correlation between GI and both RAG and RDS.[14]

Although the RAG and RDS values are not for the same food samples as those used to determine the GI values, the data suggest that RAG values and the magnitude of the glycaemic response will be closely correlated. GI values are expressed on the basis of a common amount of available carbohydrate and are normalised to the glycaemic response to a standard test meal (glucose or white bread).

In order to use GI values to make comparisons between foods it is necessary to know the amount of available carbohydrate for each food, since the GI is essentially a measure of the proportion of this that is digested and absorbed rapidly. Thus, a food with a low content of available carbohydrate will nonetheless have a high GI value if a substantial proportion of this is digested and absorbed rapidly. RAG values, by contrast, are expressed as g per 100g of food as eaten and allow direct comparisons for equal weights of foods, and make no assumption about available carbohydrate content. Although only starchy foods were considered here, GI and RAG values are not limited by the source of the rapidly available glucose and are measurable for all foods.

7 RESISTANT STARCH IS NOT FIBRE

Johnson and Southgate state that the, largely proven, dietary fibre hypothesis is:[19]

> *"Diets rich in foods containing plant cell wall material in a relatively natural state are protective against a range of diseases that are prevalent in Western affluent communities."*

In considering material to measure as dietary fibre, Johnson and Southgate state that:[19]

> *"Dietary fibre is derived from the plant cell walls in foods."*

and that measurement of dietary fibre requires techniques that provide an index of the plant cell walls in foods. Those authors suggest that because the non-starch polysaccharides

Table 4 *The 39 Starchy Foods are Ranked by RAG and Shown with GI Values Taken from the Literature*

Food	RAG/ (g/100g)	GI	Food	RAG/ (g/100g)	GI
Haricot bean	4	45	Sweet corn	16	87
Kidney bean	5	42	Parboiled rice	17	67
Chickpea	5	52	White rice	17	83
Frozen pea	5	74	Rye wholemeal bread	25	58
Beans in tomato sauce	6	60	Oat bran	32	84
Red lentil	8	42	Wholemeal bread	33	99
Marrowfat pea	8	68	All Bran	33	73
Potato (new)	8	101	White bread	38	100
Pearled barley	8	31	Digestive biscuit	40	82
Kidney bean (canned)	8	74	Porridge Oats	42	71
Chickpea (canned)	10	60	Potato crisps	43	74
Butter bean	10	52	Rich tea biscuit	48	80
Pinto bean	10	60	Oatmeal biscuit	49	78
Sweet potato	10	70	Shredded Wheat	50	97
Buckwheat	12	74	Ryvita crispbread	50	95
Instant potato	12	116	Weetabix	59	109
Macaroni	14	64	Puffed Wheat	64	110
White spaghetti	14	67	Water biscuit	66	91
Brown rice	15	96	Rice Krispies	73	117
Yam	16	74			

Correlation of Carbohydrate Fractions with Glycaemic Index

Carbohydrate Fraction	Cereals (n 23)	Other (n 16)	All (n 39)
Total carbohydrate	NS	NS	NS
Total starch	NS	NS	NS
Slowly Digestible Starch	−0.703***	−0.622**	−0.663***
Resistant Starch	−0.487**	−0.602**	−0.621***
Rapidly Available Glucose	0.756***	0.696**	0.761***
Rapidly Digestible Starch	0.727***	0.660**	0.719***
Total NSP	NS	NS	−0.406**
Soluble NSP	NS	NS	−0.442**
Insoluble NSP	NS	NS	−0.338*

P values: * <0.05; ** <0.01; *** <0.001

(NSP) account for about 90% of most plant cell walls, the measurement of dietary fibre as the NSP in plant foods is a good marker for the high-fibre diet for which benefit to health has been shown. It has been suggested by some to define dietary fibre on the grounds of indigestibility in the small intestine. This would allow the inclusion of non-cell wall material such as fructo-oligosaccharides and synthetic materials like Polydextrose, as well as retrograded starch, Maillard reaction products and other substances formed as the result of food processing. The current promotion of products free of plant cell-walls as a source of "high-fibre foods" emphasises the importance of adhering to the original concept of fibre if the consumer is not to be misled.

In considering novel polysaccharides, Johnson and Southgate state:[19]

"If, as some would advocate, the definition of dietary fibre is simplified to that of indigestibility alone, then there are logical arguments for including all indigestible components of the diet. However, this approach is at variance with the original concept of dietary fibre, and in effect would entail a new hypothesis that the amount of indigestible matter in a diet, regardless of its origin or composition, is responsible for protective effects. There is at present no evidence for such a hypothesis."

This is in agreement with Hugh Trowell, one of the authors of the fibre hypothesis, who offered the following definitions of dietary fibre:

(1) *"the skeletal remains of plant cell walls"* (the source definition) [20]

(2) *"derived from plant cell walls and not digested by human alimentary enzymes, starch was excluded from all these definitions and was never named at that time or subsequently as a constituent of dietary fibre"* [21]

Dietary fibre may affect the rate and extent of starch digestion in the small intestine but, as pointed out by Trowell [20,21] and by Johnson & Southgate,[19] starch is totally divorced from the dietary fibre concept. In accordance with this the EC Scientific Committee for Food stated:[22]

"There was lengthy discussion as to the material that should be defined as fibre for the purposes of nutrition labelling. In particular the inclusion or not of Resistant Starch in the definition of fibre was argued extensively. The Committee decided that the material to be considered as fibre for the purposes of nutrition labelling should be confined to non-starch polysaccharides of cell-wall origin."

If resistant starch was included (as it is, in part, by the AOAC Prosky procedure)[23], dietary fibre values would be highly dependent on food processing and divorced from the fibre hypothesis. Such values would not be a marker for the high-fibre diet rich in fruits, vegetables and whole-grain cereal products for which health benefits have been shown and which is recommended in the national dietary guidelines. The consequence of including starch and other non-plant-cell-wall material would be that dietary fibre would cease to be a meaningful term. Recent studies suggest that the inclusion of RS could lead to dietary fibre being a measure of substances detrimental to health (see later).

8 EFFECTS OF FOOD PROCESSING ON THE RATE OF STARCH DIGESTION

Food processing for the pre-agriculture diet consists of little other than removal of inedible or unpalatable parts. Some of the foods are cooked but most fruits and some vegetables are consumed raw. The structure of plant foods in that type of diet is largely intact and the beneficial effects of the encapsulation of nutrients, including sugars and starch, has been discussed above.

Modern food processing techniques, especially milling, may destroy the plant cell structure, and starch may be completely gelatinised. This can lead to rapid digestion of starch in the small intestine. The starch in wheat flour is 48% RDS and 49% SDS as measured *in vitro*. The ratio is changed from 1:1 in the flour to 10:1 in bread, where the values are 90% RDS and 10% SDS because the starch granules are gelatinised during the bread-making process. The result is a substantial increase in the RAG value.

There are good reasons not to increase the RAG values of foods (see later) and the desirability of increasing RS intakes has yet to be demonstrated. It is possible to manipulate the digestibility of starch in foods by the use of appropriate processing techniques. The objective for the food industry should be to develop products with an increased proportion of starch that is slowly digestible, resulting in moderate glucose and insulin responses, and this may be achieved in two principal ways: (1) Preservation of the cell-wall structure plant (dietary fibre) of plant foods will slow the rate at which starch is digested in the small intestine. (2) Introduction of secondary structure that will hinder access of amylase, as in pasta, can result in starch being slowly digested.

Table 5 *Formation of "High-fiber White Flour" by Repeated Heating and Cooling. Values are Mean (SD) for Five Bread Flours and Five Pastry Flours. Data from Ranhotra et al.*[24]

	AOAC 'Fibre' /g/100g DM	
	Bread Flours	*Pastry Flours*
Untreated	2.7 (0.1)	2.4 (0.1)
Heated/cooled	11.2 (0.6)	12.4 (1.3)

9 NEW PRODUCTS THAT MAY MISLEAD THE CONSUMER

Starch can be made resistant to digestion by heating and cooling (Table 5[24]) but this may not be desirable (see Discussion).[25,26] Table 5 shows the massive increases in AOAC Prosky "Fibre" values that can be achieved by simple heating and cooling of wheat flour. Resistant starch products are being advertised as "A new tool for creating fiber-rich foods". These products measure as much as 35% "Fiber" by the AOAC Prosky procedure but contain no plant cell-wall material and thus yield a value of zero for NSP. Apart from fermentation in the large intestine, resistant starch shares none of the properties traditionally associated with dietary fibre. The use by the food industry of high-RS material in the production of snack foods or breakfast cereals can result in apparently dramatic increases in "Fiber" content if the values are obtained by the AOAC Prosky procedure.

There are two important reasons why the marketing of such products as "fiber-rich" should be viewed with caution:

(1) the consumer will be seriously misled in the choice of foods that comprise a truly high-fibre diet and will not enjoy the health benefits associated with a diet rich in unrefined plant foods;

(2) the consumption of increased amounts of RS has been postulated to be of potential benefit to health but there is no evidence to support this claim. In fact, recent results from animal studies have suggested that high-level intakes of RS may actually be detrimental to health.[25, 26]

10 DISCUSSION

There are significant differences between the lifestyle and diet of modern Western societies and their Hunter-Gatherer ancestors. The cereal-based Western diet is highly processed and is energy-dense. National dietary guidelines recognise some of the potential hazards of the modern diet and recommend an increased consumption of "complex carbohydrates". The guidelines encourage the consumption of a range of fruits, vegetables and whole-grain cereal products as the natural sources of dietary fibre. The benefits of a naturally high-fibre diet are well documented but the consequences of increased starch consumption are less clear. It is important to recognise that not all dietary starch is digested at the same site or rate, or to the same extent.

Cereal foods tend to be highly milled, which releases nutrients, including starch, from within the plant cells. The starch in many modern processed foods is fully gelatinised and is digested rapidly in the human small intestine. Most modern people are relatively sedentary, and the combination of inactivity and ready access to energy-rich foods may lead to obesity and insulin resistance, which in turn may result in an inability to cope with modern dietary carbohydrate.[27-29]

Many modern processed foods are highly refined, and a vast range of nutrient and non-nutrient food additives are used to improve the organoleptic and storage properties of starchy foods. The food industry has the capacity to process enormous quantities of foods and consumers are prepared to pay a high price for what they see as desirable foods. The consumer does not need advice with respect to the organoleptic properties of foods. However, the situation is very different with respect to the nutritional value of foods. It is not possible to identify the origin of many modern foods by sight and the consumer must rely totally on the information provided by the food label and by health bodies.

To ensure public health and avoid soaring national health budgets, it is of the utmost importance that food labelling and nutritional advice on dietary carbohydrates, including dietary fibre, are scientifically sound. In accordance with this, the national guidelines advocate the consumption of a naturally high-fibre diet, for which benefits to health are known, and discourage the use of so-called fibre supplements other than for pharmaceutical purposes. Food labelling for Fibre as naturally occurring plant cell-wall NSP aids the consumer in choosing the recommended diet.

National guidelines recommend a decrease in fat intake and that the shortfall in energy should be compensated for by an increased intake of starch. In light of present knowledge, it is important to consider the rate of starch digestion in the human small intestine. In 1977 Heaton, in what is now recognised as a classic study, demonstrated that blood glucose and insulin responses were correlated with the extent of destruction of plant cells by processing.[30] Since then, the effect of physical structure has been shown for a range of foods, including wheat, rice and potato.[31-35] Consumption of foods with high RAG values (Table 3) is likely to lead to elevated plasma glucose levels and consequent high levels of insulin production. High levels of circulating glucose are associated with the onset and control of diabetes and with increased non-enzymatic protein glycation, which is implicated in the ageing process.[36] Hyperinsulinaemia stimulates the synthesis of cholesterol, and is associated with tumorigenesis and increased tumour growth in a number of cancers, as well as with the onset of diabetes.[3,37] Increasing knowledge about the undesirable effects of high circulating levels of glucose and insulin is prompting health workers to formulate a demand for foods with low levels of rapidly available carbohydrate. It is very important that increased starch intakes are in the form of slowly digestible starch (SDS).

Resistant starch (RS) is starch that escapes digestion and absorption in the human small intestine. Normal food processing results in only small amounts of RS. The UK diet provides an average of about 3 g of RS per capita per day and there is no evidence that these modest amounts of RS are detrimental to health.However, certain types of processing and the addition of, for example, high-amylose corn starch may result in substantial amounts of RS in foods. Fermentation of RS in the human large intestine has been shown to reduce faecal ammonia, which is potentially beneficial to health.[38] The short-chain fatty acids produced by fermentation of carbohydrates, including RS, have been implicated as a protective factor against colon cancer.[39] However, recent studies have shown that RS can enhance tumour formation in rats,[25] and in a mouse model.[26] Although detrimental effects have been shown only in animal studies, the desirability of increasing levels of RS in foods requires further evaluation.

The fact that many modern starchy foods are not fully compatible with health is not an argument for resorting to a Hunter-Gatherer diet. The new knowledge of the importance of dietary carbohydrate presents the food industry with a tremendous opportunity to contribute to public health by restricting the use of the term dietary fibre to naturally occurring plant cell walls and by the development and marketing of starchy foods that are compatible with health.

References

1. K. O'Dea, 'Western Diseases: Their Dietary Prevention and Reversibility' (N. J. Temple and D. P. Burkitt eds), Humana, Totawa NJ, 1994, p. 349.
2. G. McKeown-Eyssen, *Cancer Epidemiol. Biomarkers Prevent.*, 1994, **3**, 687.
3. E. Giovannucci, *Cancer Causes and Control*, 1995, **6**, 164.
4. H. S. Wasan & R. A. Goodlad, *Lancet*, 1996, **348**, 319.
5. H. Vlassara & R. Bucala, *Diabetes*, 1996, **45 (suppl. 3)**, S65.
6. N. J. Temple and D. P. Burkitt, editors of 'Western Diseases: Their Dietary

Prevention and Reversibility', Humana, Totawa NJ, 1994.
7.   H. N. Englyst and J. H. Cummings, *Am. J. Clin. Nutr.*, 1985, **42**, 778.
8.   H. N. Englyst and J. H. Cummings, *Am. J. Clin. Nutr.*, 1986, **44**, 42.
9.   H. N. Englyst and J. H. Cummings, *Am. J. Clin. Nutr.*, 1987, **45**, 423.
10.   H. N. Englyst and S. M. Kingman, 'Dietary Fiber. Chemistry, Physiology, and Health Effects' (D. Kritchevsky, C. Bonfield and J. W. Anderson eds), Plenum Press, New York, 1990, p. 49.
11.   H. N. Englyst, S. M. Kingman and J. H. Cummings, *Eur. J. Clin. Nutr.*, 1992, **46** (**suppl. 2**), S33.
12.   S. M. Kingman and H. N. Englyst, *Food Chem.*, 1994, **49**, 181.
13.   K. R. Silvester, H. N. Englyst and J. H. Cummings, *Am. J. Clin. Nutr.*, 1995, **62**, 403.
14.   H. N. Englyst, J. Veenstra and G. J. Hudson, *Br. J. Nutr.* 1996, **75**, 327.
15.   H. N. Englyst, S. M. Kingman and J. H. Cummings 'Plant Polymeric Carbohydrates' (F. Meuser, D. J. Manners and W. Seibel eds), Royal Society of Chemistry, Cambridge, 1993, p. 137.
16.   Department of Health Report on Health and Social Subjects 46. 'Nutritional Aspects of Cardiovascular Disease', HMSO, London, 1994.
17.   D. J. A. Jenkins, T. M. S. Wolever, A. L. Jenkins, R. G. Josse & G. S. Wong, *Lancet*, 1984, **ii**, 388.
18.   D. J. A. Jenkins, T. M. S. Wolever, A. L. Jenkins, L. U. Thompson, A. V. Rao and T. Francis, 'Dietary Fibre. Basic and Clinical Aspects' (G. V. Vahouny & D. Kritchevsky eds), Plenum Press, New York, 1986, p. 167.
19.   I. T. Johnson and D. A. T. Southgate, 'Dietary Fibre and Related Substances', Chapman and Hall, London, 1993.
20.   H. Trowell, *Am. J. Clin. Nutr.*, 1972, **25**, 926.
21.   H. Trowell, D. Burkitt and K. Heaton, editors of 'Dietary Fibre, Fibre-depleted Foods and Disease', Academic Press, London, 1985.
22.   Commission of the European Communities, Directorate Industry, III/E/1. Scientific Committee for Food, Document III/5481/93–EN–FINAL, Brussels, 21 October 1993.
23.   Official Methods of Analysis, 15th Edit., vol. II, section 985.29, AOAC, Arlington VA, 1990, p. 1105.
24.   G. S. Ranhotra, J. A. Gelroth and G. J. Eisenbraun, *Cereal Chem.*, 1991, **68**, 432.
25.   G. P. Young, A. McIntyre, V. Albert, M. Folino, J. G. Muir and P. R. Gibson, *Gastroenterology*, 1996, **110**, 508.
26.   J. Burn, A. Katheuser, R. Fodde, J. Coaker, P. D. Chapman, and J. C. Mathers (1996). *Eur. J. Hum. Genet.*, **4 (suppl. 1)**, 13.
27.   E. S. Horton, *Diabetes Metab. Rev.*, 1986, **2**, 1.
28.   S. P. Helmrich, D. R. Ragland, R. W. Leung and R. S. Paffenbarger, *N. Engl. J. Med.*, 1991, **325**, 147.
29.   J. E. Manson, E. B. Rimm and M. J. Stampfer et al., *Lancet*, 1991, **338**, 774.
30.   G. B. Haber, K. W. Heaton, D. Murphy and L. F. Burroughs, *Lancet*, 1977, **i**, 679.
31.   K. Hermansen, O. Rasmussen, J. Arnfred, E. Winther and O. Schmitz, *Diabetalogia*, 1986, **29**, 358.
32.   J. Arends, K. Ahrens, D. Lübke and B. Willms, *Klin. Wochenschr.* 1987, **65**, 469.
33.   K. W. Heaton, S. N. Marcus, P. M. Emmett and C. H. Bolton, *Am. J. Clin. Nutr.*, 1988, **47**, 675.
34.   P. A. Crapo and R. R. Henry, *Am. J. Clin. Nutr.*, 1988, **48**, 560.
35.   F. R. J. Bornet, D. Cloarec, J.-L. Barry, P. Colonna, S. Gouilloud, J. Delort Leval and J.-P. Galmiche, *Am. J. Clin. Nutr.*, 1990, **51**, 421.
36.   D. G. Dyer, J. A. Dunn, S. R. Thorpe, K. E. Bailie, D. R. McCance and J. W. Baynes, *J. Clin. Invest.*, 1993, **91**, 2463.
37.   K. W. Heaton, 'Western Diseases: Their Dietary Prevention and Reversibility'

(N. J. Temple and D. P. Burkitt eds), Humana, Totawa NJ, 1994, p. 187.

38. A. Birkett, J. Muir, J. Phillips, G. Jones & K. O'Dea, *Am. J. Clin. Nutr.,* 1996, **63**, 766.

39. I. P. van Munster, A. Tangerman & F. M. Nagengast, *Digest. Dis. Sci.,* 1994, **39**, 834.

DENTAL EFFECTS OF STARCH

T. H. Grenby

Department of Oral Medicine and Pathology, UMDS, Guy's Hospital,
London, SE1 9RT, United Kingdom

1 INTRODUCTION

At the start of the 1960's virtually no information existed specifically on the dental effects of starch. All carbohydrates were grouped together as having the same nutritional and physiological properties. In 1960 a programme of research was initiated at the Research Association of British Flour Millers, St. Albans, to investigate whether this was correct. Most of the research on the influence of starches on dental health was carried out in the next 30 years. This paper reviews the main findings that emerged.

Dental caries levels were very high in the early 1960's, and there was serious concern over the influence of different constituents of the diet, particularly the carbohydrates, on the prevalence of tooth decay.[15,18] Surveys such as these did not adequately distinguish between different types of carbohydrates. Data from four main experimental approaches were therefore employed to try to differentiate between food carbohydrates and to investigate the cariogenicity (caries-promoting capacity) of sugars and starches:

(a) Epidemiological surveys of population groups, attempting to relate their dental condition to their diets and eating patterns.

(b) Clinical studies of factors believed to be important in the dental caries process in man, relating them to specific constituents of the diet.

(c) Evaluation of the cariogenicity of purpose-blended diets in specially-bred strains of caries-active laboratory rats under standardised conditions.

(d) Experiments *in vitro* measuring factors that play a part in the caries process, such as acid development, dental plaque (polysaccharide) formation, microbial growth and the attack on dental enamel.

Most of the research on the dental properties of starches has been carried out by methods (c) and (d). Surveys (a) would not be capable of identifying the specific dental effect of just one single item of the diet, and clinical studies (b) are expensive to set up, require ethical approval, and of course cannot be used to study the degenerative disease of dental caries directly, so that investigations have to be confined to some factor associated with caries.

2 COMPARISON OF SUGARS AND STARCHES

In some of the earliest work on this the findings were very clear: sugars (sucrose and glucose) in the diet of caries-active laboratory rats were cariogenic, whereas the level of caries was very low on raw wheat starch.[6] This was confirmed in later experiments and by other workers.[5]

Caries activity levels in seven generations of rats fed on sugar or cornstarch diets for 84 or 150 days were measured.[21] A much higher incidence of occlusal caries on sucrose and glucose diets than on cornstarch was observed, but the smooth-surface caries scores did not show any consistent difference. One factor that may have played a part in this is the fineness of the cornstarch particles, as it has been said that coarse corn particles may be hard enough to initiate lesions by contributing to fracturing of the teeth.

It has proved hard to extrapolate these findings to man, since a normal mixed diet of course contains both sugars and a variety of starches, but in a survey of 405 children, Rugg-Gunn et al.,[20] observed a trend towards a stronger relationship between sugar intake and dental caries than between starch intake and caries.

3 EFFECT OF COOKING ON STARCH

Of course cooking and processing can be carried out in various ways and under a range of conditions, so that the products can vary widely in the degree of denaturation, gelatinization and retrogradation. In one of the first comparisons, in which 5 parts wheat starch were cooked with 4 parts water in a steam oven at 100°C for 90 min, then dried and pulverised in a hammer-mill, the starch was non-cariogenic, no matter whether cooked or raw. On 66% starch regimes, the caries scores were all very low, averaging 2.1 for raw starch and 2.6 for cooked starch, but were not significantly apart.[7]

A few years later Frostell & Baer[4] recorded the caries scores in groups of 23 or 24 rats fed on 66% starch diets for 30 days. Figures for four types of unmodified starch were uniformly low, whereas pregelatinized arrowroot, potato, amioca, tapioca and wheat starches all appeared to be more cariogenic. Their caries potential was presumably related to the pregelatinization process they had undergone. The authors believed that gelatinization liberates more amylopectin than amylose, and that amylopectin is the more cariogenic of the two molecules.

One of the salivary hydrolysis products of starch is maltose. Brudevold et al.[1] noted that cooked starch produced higher salivary maltose concentrations than raw starch, and raw starch had no demineralizing potential. The demineralizing power of cooked starch was related to its rapid hydrolysis but not to its clearance time from the mouth. In later work measuring dental enamel permeability, it was found that the higher the concentration of gelatinized starch in the mouth, the longer the clearance time, and it was concluded that the starch in baked or cooked foods may have significant demineralization potential.[2]

Heating dry wheat starch alters its molecular structure so that it can be readily metabolised by Streptococcus mutans, one of the oral micro-organisms that participates in the caries process.[22] Some further views on cooked starch were given by Firestone et al.,[3] whose time-controlled administration of carbohydrate diets to laboratory rats indicated that cooked wheat starch was cariogenic but less so than sucrose.

4 DENTAL PROPERTIES OF HIGH-STARCH FOODS

Naturally, interest in the dental properties of pure starches has been very limited. Of greater practical significance are the effects of high-starch foods which are dietary staples in various parts of the world, e.g. bread, rice, potatoes, pasta and other cereal foods.

There has been an unconfirmed assumption that because they are initially low in sugars, they will not be cariogenic, but of course the possibility exists that cooking or gelatinizing the starch can alter this (see preceding section).

5 EXPERIMENTS ON BREAD

One of the first high-starch staples to be investigated for its cariogenicity was bread. White and wholemeal four and bread were evaluated at a level of 66% in the diets of caries-active laboratory rats, and then finally large groups were fed on 100% dried bread diets for 11 weeks from weaning.[8] The incidence of caries was related to the sucrose content of the diet and not to the proportion of flour or bread in it. The levels of caries were extremely low on both types of bread, with no significant differences observed between white and wholemeal. The findings on the low cariogenicity of flour in the absence of sucrose were corroborated by König & Grenby.[16]

6 BISCUITS

Biscuits (cookies), including crackers, wafers and crispbreads, with levels of starch generally in the range of 40 to 70%, represent another category of high-starch foods on which dental information was lacking. In the first series of experiments, evaluating the effects of sweet biscuits (approx. 40% sugars) in two strains of caries-active laboratory rats, the biscuits emerged as highly cariogenic, whereas control diets containing the same amounts of sucrose and starch but not baked into biscuits, were significantly less cariogenic.[13]

These findings were confirmed in later experiments, and it was also shown that digestive biscuits containing 48% starch and with alternative sweeteners replacing sucrose were significantly less cariogenic than conventional digestive biscuits containing the same level of starch (48%) but sweetened as normal with 15% of sucrose.[11]

7 CHILDREN'S RUSKS

When six different types of cereal-based rusks, total carbohydrate contents 69-79% were studied in laboratory rates and *in vitro*,[14] their cariogenicity generally correlated well with their sucrose content, but it was observed that cereal components in the rusks could play a part in governing their adhesiveness to the tooth surface and their fermentability. The dental health implications of these findings are (a) that the more adhesive carbohydrate foods are to the teeth, the longer the contact time for acids derived or formed from the foods to advance the caries process; and (b) the greater their fermentability, the higher the risk of breakdown of short-chain carbohydrates that can be metabolised by bacterial or salivary enzymes to acids that can attack the mineral matter of the teeth, as in caries. A sucrose-free rusk formulation produced the lowest caries levels of all.

8   BREAKFAST CEREALS

These are another example of high-starch cereal products. The chief feature of interest in relation to their starch content when their dental properties were studied in caries-active rats was the low caries score on a diet formulated with a whole-wheat cereal (81.3% carbohydrate) and zero sucrose.[12] This was a further indication that processed or cooked starch products in the absence of sucrose have a comparatively low caries potential, but the study went on to show that the inclusion of as little as 9.5% of sucrose in a presweetened breakfast cereal increased caries levels by a highly significant margin.

9 HIGH-STARCH SNACK FOODS

Many very popular types of savoury snack foods are also high in starches. Particular use is made of cereal and potato starches which are processed in various ways, sometimes under highly disruptive conditions. Research on their dental properties in laboratory animals has shown wide variations in their cariogenic potential, depending partly on the severity of the processing the starch has undergone.[9,10] In some cases the caries levels associated with the savoury snacks were not far below those from semi-sweet biscuits.

One stage in the production of the snacks is extrusion cooking, in which a slurry of starch-rich cereal or potato is subjected to conditions of high temperature, high pressure, high moisture and high shear forces. More highly processed varieties of snacks are double-extruded, gelatinized, expanded and fried. The inference is that all this can increase the susceptibility of the starch to attack by salivary or microbial α-amylase in the mouth. This effect has been studied *in vitro* , and a rise in free maltose, which can be metabolised to cariogenic acids by oral micro-organisms, has been demonstrated.[10]

Intra-oral plaque pH changes were found to be slower after starch snacks than after the consumption of sucrose.[19]   Finally, Lingström et al.,[17] observed that all processing methods, including extrusion cooking, drum-drying, popping, bread making and steam flaking, raised the fermentability of starch in human dental plaque, and commented that gelatinization was important in determining the extent of hydrolysis of the starch by α-amylase.

10 CONCLUSIONS

It appears from the evidence collected over a period of about 30 years that:-

(a) Starch in its natural state is not cariogenic or of very low cariogenicity.

(b) But starch is not consumed in this form in Western diets. Cooking or processing denature it and raise its potential cariogenicity.

(c) Mixtures of processed starches and sugars are more highly cariogenic than starches alone, providing the substrate for oral micro-organisms to generate acids that can attack dental enamel.

(d) Dental caries attack may also be promoted by processed starches contributing to the persistence or adherence of mixtures of fermentable foods at susceptible sites on the teeth.

REFERENCES

1. F. Brudevold, D. Goulet, A. Tehrani, F. Attarzadeh and J van Houte, *Caries Res.,* 1985, **19**, 136.
2. F. Brudevold, D. Goulet, F. Attarzadeh and A. Tehrani, *Caries Res.,* 1988, **22**, 204.
3. A. R. Firestone, R. Schmid and H. R. Mühlemann, *Archs. Oral Biol.,* 1982, **27**, 759.
4. G. Frostell and P. N. Baer, *Acta Odont. Scand.,* 1971, **29**, 401.
5. R. M. Green and R. L. Hartles, *Brit. J. Nutrition,* 1967, **21**, 921.
6. T. H. Grenby, *Archs. Oral Biol.,* 1963, **8**, 27.
7. T. H. Grenby, *Archs. Oral Biol.,* 1965, **10**, 433.
8. T. H. Grenby, *Brit. Dent. J.,* 1966, **121**, 26.
9. T. H. Grenby, *Brit. Dent. J.,* 1990, **168**, 353.
10. T. H. Grenby, *Chem. Brit.,* 1991, **27**, 638.
11. T. H. Grenby and J. M. Bull, Royal Society 'Developments in Food Carbohydrate I' (Ed. G. G. Birch and R. S. Shallenberger), Applied Science Publishers, London, 1977, 169.
12. T. H. Grenby and J. M. Bull, *Archs. Oral Biol.,* 1978, **23**, 675.
13. T. H. Grenby and F. M. Paterson, *Brit. J. Nutr.,* 1972, **27**, 195.
14. T. H. Grenby, A. Phillips and M. G. Saldanha, *Brit. Dent. J.,* 1989, **166**, 157.
15. P. J. Holloway, P. M. C. James and G. L. Slack, *Brit. Dent. J.,* 1963, **115**, 19.
16. K. G. König and T. H. Grenby, *Archs. Oral Biol.,* 1965, **10**, 143.
17. P. Lingström, J. Holm, D. Birkhed and I Björck, *Scand. J. Dent. Res.,* 1989, **97**, 392.
18. W. D. McHugh, J. D. McEwen and A. D. Hitchin, *Brit. Dent. J.,* 1964, **117**, 246.
19. K. K. Park, B. R. Schemehorn, J. W. Bolton and G. K. Stookey, *Amer. J. Dentistry,* 1990, **3**, 185.
20. A. J. Rugg-Gunn, A. F. Hackett and D. R. Appleton, *Caries Res.,* 1987, **21**, 464.
21. J. H. Shaw and J. K. Ivimey, *J. Dent. Res.,* 1972, **51**, 1507.
22. M. E. Thomson and R. B. Wills, *Archs. Oral Biol.,* 1976, **21**, 779.

STARCH FUNCTIONALITY IN FOOD PROCESSING

Jay-lin Jane, Department of Food Science and Human Nutrition and Center for Crops Utilization Research, Iowa State University, Ames, Iowa 50011, USA

Starch, a D-glucan with α 1-4 linked backbone and α 1-6 linked branches, consists of amylopectin (branched molecules), and amylose (primarily linear molecules). The starch molecule has a coiled conformation in an aqueous solution and has a tendency to form helical structures: double or single helices (1). The glucose anhydrous unit of a starch molecule carries hydrophilic hydroxyl groups and hydrophobic hydrogens at the equatorial and the axial positions of the C1 sugar ring, respectively. Starch is produced in granules by higher plants for energy storage. In the granule, starch molecules are present as a semicrystalline double helical structure (2), (3). After dispersion in an aqueous medium, starch molecules form helical complexes if chemicals containing hydrophobic tails, such as 1-butanol and fatty acids, are present in the media (4), (5). The helical complex formation is an instantaneous reaction through hydrophobic interaction between the hydrophobic tail of the complexing agent and the hydrocarbon of the starch. Without complexing agents, starch molecules slowly intertwine to form double helices and become retrograded and undergo syneresis. The double helical structure is a low energy stable conformation with some hydrocarbon folded inside of the double helices. The transformation of starch structures is shown in Figure 1.

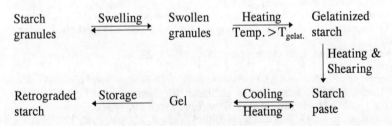

Figure 1. Transformation of starch physical structure.

The transformation of the starch structure gives versatile functions, but it also causes the instability of starch for any one selected function. For example, starch granules may lose their integrity by heating or adding chemicals; and starch pastes go through a natural aging process in which they return to a semicrystalline structure and lose their water-binding capacity. Being produced in granules, starch has the great advantage of easy isolation and processing which includes chemical modifications and subsequent washing to remove by-product residues. Because of these advantages, starch can be produced as one of the most economical commodities. Functions of starch applied for food processing can be discussed in groups according to the physical structures. Functions of starch in nonfood applications have gained significant interest because of the abundance and low cost of starch and because of growing concern for environmental pollution caused by non-biodegradable petroleum-based polymers.

1. FUNCTIONS PROVIDED AT GRANULAR FORMS

Starch is isolated as granules. The densities of starch granules are around 1.5g/cm^3 (3). This high density and the granular structure facilitate the separation of starch by centrifugation/gravity sedimentation or by filtration during isolation, chemical modifications, and washing. Starches isolated from different botanical sources display characteristic granular morphology (6). Starch granules display various shapes, including spherical, oval, polygonal, disk, elongated, and kidney shapes. Normal maize starch has spherical and polygonal shapes; potato starch has oval and spherical shapes; wheat starch has a bimodal size distribution, among them, large (A) granules have a disk shape, whereas small (B) granules have a spherical shape; diffenbachia starch has an elongated submarine shape; and almost all the legume starches have a characteristic indentation on granules of bean-like shapes. Diameters of starch granules vary from submicron, such as amaranth and small pigweed, to more than one-hundred microns, such as canna starch. Other starches, such as small wheat granules have diameters of 2-3 microns; large wheat granules, 22-36 microns; potato, 15-75 microns; maize, 5-20 microns; rice, 3-8 microns; and legume starches, 10-45 microns (6). Granular starch has found many uses in both food and nonfood products.

1.1. Food Uses

In the food industry, granular starch has been commonly used as dusting agents for candy, carrying agents for baking powder, and starch molds for gum drop manufacture. Small granular or small particle starch (7) and microcrystalline starch (8) with diameters ranging from submicrons to 2 microns are proposed for fat substitutes or fat mimetics. Mixtures of small, soft starch particles dispersed in a starch gel matrix resemble the texture of butter in which fat micelles of diameters ranging from 1 to 2 microns are dispersed in liquid fat matrix (8).

Granular raw starches have demonstrated different digestibilities (9). The digestibility of starch affects the nutritional (caloric) value of raw starch present in grain or tuber for food or feed uses. Studies have shown that starches with B- and C-type X-ray crystalline patterns, such as potato starch and banana starch, suffer poor enzyme digestibility compared with A-type starch, such as maize, wheat, and rice starch (9). Williamson et al, (10) studied the digestibilities of A-type and B-type spherulites made from maltodextrins and reported the same results. The packing of the double-helices in the crystalline structure of B-starch has an orthogonal unit cell, consisting of an open channel at the center, whereas the A-starch has a hexagonal unit cell structure (3). Reasons for B-starch being more resistant to enzymatic and acidic hydrolysis is not clear. Most native B-starch has longer branch chains (11); it is plausible that the long chain double helices may affect the digestibility. It is also possible that B-starch has branch points present within the larger crystalline region and, thus, is more resistant to enzyme hydrolysis (12).

1.2. Nonfood uses

Starch has gained increasing interest for nonfood uses. In addition to dusting agents and facial powders, granular starch has been used in various products depending on their shapes and sizes. Examples include the use of large granules of wheat starch for carbonless copy paper. The large granule wheat starch has a disk shape, smooth surface, and a proper granule size, which has been demonstrated to be desirable as "stilt material" for coatings on carbonless copy paper to protect encapsulated carbon beads until the paper is pressed (13). Flavor carriers made from spherical aggregates of starch granules have been reported by Zhao and Whistler (14). Small granules, such as amaranth, small wheat, and rice, are preferred for this application. The spherical aggregates, bound by protein and various polysaccharides, contain interconnecting cavities that provide extensive porosity. Starch granules of smaller sizes have also been found to improve the tensile properties of starch-filled polyethylene films (15), (16).

Starch of small granule size has been found desirable for many uses, as previously described. Small granule starch, however, is difficult to isolate because of its slow settling and contamination of protein residues. Traditionally, wet milling of small granule starch involves high-concentration alkaline steeping (17). The process is expensive and not environmentally friendly. A combination of low alkaline and enzymatic treatments has been demonstrated successfully for amaranth starch isolation (18). Small particle starch can also be produced from common starch, such as maize, by mild acid hydrolysis followed by mechanical attrition (7).

2. FUNCTIONS PROVIDED BY STARCH PASTES

Starch pastes are widely used in food products for thickening, fillings, oil mimetics, and providing body texture to beer and soft drinks. Predominant markets of starch are in the paper and textile industries, in which starch pastes are used as binding and sizing agents.

Properties of pastes made from starches of different botanical sources vary greatly. It is well known that potato starch produces a crystal clear paste with substantially larger viscosity compared with normal and waxy maize, wheat, rice, and tapioca starch. It is also known that waxy starch and root and tuber starches (e. g., tapioca starch) have lower pasting temperatures and greater paste clarity than normal cereal starches (e.g., wheat, normal maize, and rice). Structural analyses of these starches by using chemical and enzymatic methods (19) and P^{13}-nmr (20) have revealed that potato starch consists of substantial amounts of phosphate derivatives, whereas, normal cereal starches consist of high concentrations of phospholipids (20). The phosphate monoester groups of potato and other starches carry negative charges which repel one another and result in greater viscosity and clarity. The phospholipids in the normal cereal starches, however, complex with amylose and long-branch chains of amylopectin (21) and generate junction zones resulting in turbidity (22) and higher pasting temperature. Tapioca and most other root and tuber starches consist of low concentration phosphate derivatives and no phospholipids (20); thus, the starches have relatively high clarity and low pasting temperatures. Waxy starch contains little or no amylose and phospholipids and, thus, displays great clarity and a low pasting temperature.

Methods have been developed to improve the properties of starch pastes by chemical, physical, enzymatic, and genetic modifications. Most commonly used approaches for commercial starch products include derivatizing, cross-linking, hydrolyzing, and oxidation (23). Increasing research efforts have focused on genetic modification by breeding and genetic engineering to produce starch of modified properties (24).

2.1. Chemical modifications (23)

Chemical derivatives, such as phosphate, succinate, hydroxypropylate, and acetate, are commonly added to starch to control retrogradation and to increase the viscosity and smoothness of starch pastes. Phosphate, succinate, and other acidic derivatives carry negative charges at neutral pH, which generate repulsion forces and result in larger viscosity, lower gelatinization temperature, and enhanced clarity. One drawback of the starch is its sensitivity to the salt content in the paste. A trace of salt can drastically decrease the viscosity of the starch paste as a result of the charges of the acidic groups being neutralized by the cations of the salt. Hydroxypropylate and acetate have both been commonly used to prepare starch derivatives, which reduce the gelatinization temperature, particularly for high-amylose starch, and enhance the stability of the starch paste. Hydroxyethylated starch is only approved for industrial uses. Starch acetate, with some hydroxyl groups esterified by acetyl groups, is more hydrophobic and has been reported to produce crispy textures (25). The glass transition temperatures (T_g) of starch phosphate, maleate, and succinate are lower than that of native starch (26), but that of starch acetate is higher than native starch (27).

Octenylsuccinate starch, carrying both a hydrophobic octenyl group and a hydrophilic carboxylic group, displays a unique property. The starch has been used as an emulsifier for beverage emulsions, salad dressings, encapsulations, etc. An aluminum complex of the octenylsuccinate granular starch, of which the carboxylic groups are chelated by aluminum ions, possesses a highly hydrophobic surface and displays water-repellent and free-flowing properties (23).

During processing, starch is subjected to many harsh conditions, such as low pH, high temperature, mechanical shearing and pumping. To prevent loss of viscosity by hydrolysis or degradation of starch, starch is commonly cross-linked (23). Starch phosphate diester and starch adipate diester are the two most common cross-linked starches for food uses. Other cross-linked starches, such as those prepared by epichlorohydrine and aldehydes, are not approved for food uses.

Combinations of chemical derivatives and cross-linking have been demonstrated to provide stability toward acidic, thermal, and mechanical degradation of starch and to prevent retrogradation during storage. Dual modified starch, such as cross-linked hydroxypropylate starch, is particularly desirable for frozen desserts and other food products to maintain good qualities during processing and storage (23).

Converted starch is produced by acid hydrolysis or oxidation of starch. The converted starch has reduced paste viscosity and can prevent gelling. Thus, pastes of increased concentration can be prepared and used for various food and nonfood applications (such as adhesives for stamps).

2.2. Enzymatic modifications of starch

Maltodextrins can be prepared by partial hydrolysis of starch by using acid and/or enzymes (28). Maltodextrins of different molecular sizes, indicated as dextrose equivalent (DE), are useful for adding body texture to beer and soft drinks, for additives to decrease the glass transition temperature of starch, and for carriers. Corn syrups and high-fructose corn syrups are also products of enzymatic modifications of starch (28).

Cyclodextrins are prepared by enzymatic (cycloamylose glucanotransferase) reactions of starch (29), (30). Common cyclodextrins produced by the enzyme are α-, β- and γ-cyclodextrins, consisting of six, seven, and eight glucose units, respectively. Cyclodextrins larger in size (nine and more glucose units) are produced but are rare (31). Cyclodextrins have hydrophobic cavities which have affinity to complex with hydrophobic structures of many chemicals to form inclusion complexes. Cyclodextrins have been used to stabilize volatile flavor compounds, to remove unpleasant flavors from foods (such as the bitter taste of citrus juice) (32), to protect drugs from oxidation, and to achieve speedy delivery of drugs by fast absorption and increased solubility of the complex.

2.3. Physical modifications of starch (33)

Preparing starch paste by cooking may become a hurdle for some applications of starch pastes and gels, such as for instant foods and when high-amylose starch is involved. Because most high-amylose starch has gelatinization and pasting temperatures above 100°C (the boiling temperature of water), special cooking equipment is needed to process the starch. To solve this technical difficulty, instant starch products have been produced. Several technologies have been developed to manufacture instant starch. Conventional approaches are to precook the starch slurry, followed by drum drying and grinding (34). Products prepared by this method display low viscosities, as a result of degradation of the starch during the processing.

Many other techniques have been developed within the past decade to produce granular cold-water-soluble starch. Pitchon et al., developed a method of treating starch by using injection and nozzle-spray drying (35). Starch slurries can be atomized and cooked in a nozzle chamber for a short period and spray dried. The process produces granular precooked starch. Starch can also be thermally treated (150-170°C) (36), (37) or treated with caustic alkali (NaOH or KOH) (38) in aqueous alcoholic media or treated with hot aqueous polyhydric alcohols (39). Granular cold-water-soluble (GCWS) starch displays large viscosity, equivalent or better than the pastes prepared by freshly cooked native starch counterparts (37). The formation of GCWS starch is attributed to the gelatinization of starch, either by heat or by alkali, in media which restrict the swelling of starch. Simple alcohols, such as ethanol, methanol, or propanol, are preferred because they form helical complexes with dispersed starch molecules to stabilize the molecules and to prevent starch from retrograding (form double helical structures). On drying, the alcohol is evaporated, and the starch remains in the single helical conformation with an empty cavity while the starch is kept dry in a glassy state. When water is added to the starch, the metastable single helical starch becomes dispersed and forms a paste instantly.

Pastes made from GCWS starch which is prepared by the alcoholic-alkaline method display improved freeze-thaw stability compared with those made from native starch (40). The suppressed retrogradation rate may be attributed to conformational changes induced by the alkali treatments.

The glass transition temperature (Tg) of starch varies with many parameters, such as chemical derivatives (26), additives, and moisture content (41). Studies have shown that in addition to starch modification, the presence of additives, such as sugars, maltodextrins and glycerol, decrease the glass transition of starch (26). The concentration of moisture drastically affects the glass transition temperature (41). The glass transition temperature determines the mechanical properties at a designated temperature. With the presence of small sugars, such as glucose and fructose, the T_g' of starch paste is significantly decreased and results in rapid retrogradation even at a freezer temperature (-20°C) (42), (43). This indicates that fructose and glucose will enhance starch retrogradation and shorten the shelf life of frozen desserts unless the freezer is set at a temperature substantially below the T_g'.

2.4. Genetic modifications of starch

With health concerns over ingesting chemicals in chemically modified starches and environmental concerns over waste water treatment and by-product recovery after conducting the chemical reactions, there is a trend to modify starch properties by genetic means (44), (45). Through breeding and genetic engineering, starch structures can be altered and the properties can be improved. For example, waxy maize starch, having negligible amylose and no long B chains (B4 & B5), has less tendency to retrograde and displays improved clarity. High-amylose maize starch, consisting of high-concentration amylose and long branch-chain amylopectin, produces good films with strong tensile properties (46) but the gelatinization temperature is very high (> 100°C). Pasting properties of normal, waxy, du waxy, and ae waxy maize starches vary, and pasting curves of the starches prepared by using an amylograph are shown in Figure 2. High-amylose starch fails to show a pasting curve because it cannot be fully gelatinized and dispersed to paste by using an amylograph heating to 97°C.

With the development of genetic engineering, starch structures can soon be genetically modified by design to produce starches of tailored properties. For example, A- and B1-chains of starch may be shortened to decrease the gelatinization temperature (47) and to slow down the retrogradation, whereas, B4 and B5 chains may be added to increase the resistance to shear thinning and increase the viscosity of the paste, resembling the properties of cross-linked starches. Increase in molecular size of amylose can also slow down the retrogradation (48), (49) and improve the mechanical properties of starch pastes and gels. Reduction in lipid (particularly phospholipid) contents may increase the clarity of the starch paste (22). Once phosphate derivatives can be genetically added to starch, there is no need of conducting chemical reactions of phosphorylation.

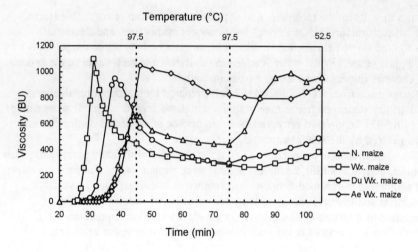

Figure 2. Amylographs of genetically modified maize starches

3. FUNCTIONS PROVIDED BY STARCH GEL

At high concentrations, starch pastes gel on cooling. Starch gels provide structures to bread, cakes, and pudding. Starch gel also gives desirable textures to foods, such as crispy coatings to fried chicken and fish (25).

To prevent the collapse of a baked cake, oxidized starch (flour) is preferred for cake batter (23). This is because the oxidized starch, prepared by treating the starch with hypochlorite or chlorine gas, carries carboxylic groups and aldehyde groups and suppresses the gelatinization temperature. The reduction of the gelatinization temperature enables starch to gelatinize and form gel in the presence of a high concentration of sugar to hold the structure of the cake. The carboxylic and aldehyde groups also interact with egg white protein through charge interactions and covalent bonds (50), respectively.

The oxidized starch is also a desirable choice for breading and battering of fried chicken and fish (25). The carboxylic and aldehyde groups of the starch enhance the binding and coating by the charge interactions and covalent bonding formed between the aldehyde groups of the starch and the hydroxy, amino, and sulfhydryl groups of meat proteins. Therefore, the oxidized starch batters provide a stable and uniform coating to the fried foods.

High-amylose starch is known for its film-formation property and higher glass transition temperature (51). Thin film formation is particularly desirable for fried food coating batters, and the thin film at its glassy state provides a crispy texture. Acetylated high-amylose starch, with higher hydrophobicity and glass transition temperature, further enhances the crispy texture of fried foods. High-amylose starch coatings have also been applied to doughnuts to reduce oil absorption. High-amylose starch produces tough gels, which are desirable for manufacturing gum drops.

4. FUNCTIONS PROVIDED BY RETROGRADED STARCH

After retrogradation takes place, starch has a semicrystalline structure. The double helical crystalline structure is highly resistant to both enzymatic and acid hydrolysis and is known as resistant starch (52). Resistant starch is normally prepared with high-amylose starch after repeated autoclaving and retrogradation and followed by enzyme hydrolysis to remove the amorphous starch (53). The resistant starch is used as a bulking agent, which generates little caloric value and has the health benefit of preventing colon cancer.

Clear noodles are made by extruding mung bean starch (containing 37% apparent amylose) paste, cooked in boiling water, and incubated at refrigerating temperature (4°C) for 24 to 48 hr to retrograde (54). The noodle is known for its low digestibility and is a desirable food for diabetic patients.

5. FUTURE DEVELOPMENTS OF STARCH APPLICATIONS

Starch has a relatively low caloric value compared with lipid and contains no cholesterol. Thus, starch meets consumers' expectations for health benefit. Dietary patterns of the mass majority will gradually shift from fat and protein to starch and complex carbohydrates. To meet the demand of high quality foods, sophisticated modifications of starch will be required. With consumers' concerns of chemically processed foods, genetic modifications of starch will be used to replace chemical means to improve starch properties. Specialty starches will be produced in the field, and some modifications of wet-milling operations will be adopted to accommodate the processing of varieties of starches.

In addition to food applications, starch will be increasingly used in nonfood applications as a biodegradable polymer and as a feedstock for chemicals. There is an increasing demand for biodegradable polymers to replace nondegradable petroleum based polymers to reduce environmental pollution. Starch is one of the most economically competitive biopolymers. Therefore, starch use will be increasing in processing aids, in biodegradable plastics and others.

REFERENCES

1. A. D. French & V. G. Murphy, Cereal Foods World, 1977, **22**, 61.
2. D. French, in Starch: Chemistry and Technology. 2nd Ed. Eds. R. L. Whistler, J. N. BeMiller and E. F. Paschall, Academic Press, Orlando, FL. 1984, P.183.
3. A. Sarko and H. C. H. Wu, Starch/Staerke, 1978. **30**, 73.
4. T. Kuge and K. Takeo, Agric. Biol. Chem., 1968, **32**, 1232.
5. Y. Yamashita, J. Polym. Sci., Part A, 1965, **3**, 3251.
6. J. Jane, T. Kasemsuwan, S. Leas, H.F. Zobel, and J. F. Robyt, Starch/Staerke, 1994, **46**, 121.
7. J. Jane, L. Shen, L. Wang, and C.C. Maningat, Cereal Chem., 1992, **69**, 280.
8. R. C. Chiou, C. C. Brown, J. A. Little, A. H. Young, R. V. Schanefilt, D. W. Harris, K. D. Stanley, H. D. Coontz, C. J. Hamdan, J. A. Wolf-Rueff, L. A. Slowinski, K. R. Anderson, W. F. Lehnhardt, and Z. J. Witczak, EP Patent 0,443,844 A1, 1991.

9. H. Fuwa, T. Takaya, and Y. Sugimoto, in Mechanisms of saccharide polymerization and depolymerization. Ed. J. J. Marshall, Academic Press, New York, NY, 1980.
10. G. Williamson, J. J. Belshaw, D. J. Self, T. R. Noel, S. G. Ring, P. Cairns, V. J. Morris, S. A. Clark, and M. L. Parker, Carbohydr. Polym., 1992, **18**, 179.
11. S. Hizukuri, T. Kaneko and Y. Takeda, Biochim. Biophys. Acta, 1983, **760**, 188.
12. P. Colonna, V. Planchot, and A. Buleon, in Starch Structure and Function Conference Proceedings. Cambridge, UK, 1996.
13. W. Nachtergaele and J. Van Nuffel, Starch/Staerke, 1989, **41**, 386.
14. J. Zhao and R. L. Whistler, Food Technol., 1994, 104.
15. S. Lim, J. Jane, S. Rajagopalan and P. A. Seib, Biotechnol. Prog., 1992, **8**, 51.
16. G. J. L. Griffin, in Wheat is unique. Ed. Y. Pomeranz, American Association of Cereal Chemists, St. Paul, MN, 1990, p.695.
17. J. Uriyapongson and P. Rayas-Duarte, Cereal Chem., 1994, **71**, 571.
18. M. Radosavjevic, J. Jane, and L. A. Johnson, Unpublished data.
19. S. Hizukuri, S. Tabata, and Z. Nikuni, Staerke, 1970, **22**, 338.
20. S. Lim, T. Kasemsuwan, and J. Jane, Cereal Chem., 1994, **71**, 488.
21. L. V. Batres and P. J. White, J. Am. Oil Chemist Soc., 1986, **63**, 1537.
22. S. A. S. Craig, C. C. Maningat, P. A. Seib, and R. C. Hoseney, Cereal Chem., 1989, **66**, 173.
23. O. B. Wurzburg Ed., Modified starches: properties and uses, CRC Press, Boca Raton, FL, 1986.
24. J. Preiss and M. Sivak, in Photoassimilate distribution in plants and Crops, Sec. A, Eds. E. Zamski and A. A, Schaffer, Marcel Dekker, New York, NY, 1995, 63.
25. S. Hegenbart, Food Product Design, January 1996, 23.
26. S. Lim, and J. Jane, J. Environm. Biodegrad. Polym., 1994, **2**, 111.
27. R. L. Shogren, Carbohydr. Polym. in press.
28. J. F. Robyt, in Starch: Chemistry and Technology. 2nd Ed. Eds. R. L. Whistler, J. N. BeMiller and E. F. Paschall, Academic Press, Orlando, FL. 1984, P.87.
29. K. Kainuma, in Starch: Chemistry and Technology. 2nd Ed. Eds. R. L. Whistler, J. N. BeMiller and E. F. Paschall, Academic Press, Orlando, FL. 1984, P.125.
30. J. Szejtli, in Biotechnology of amylodextrin oligosaccharides. Ed. R. B. Friedman. American Chemical Society, Washington, DC, 1991, p.2.
31. A. O. Pulley and D. French, Biochem. Biophys. Res. Communic., 1961, **5**, 11.
32. W. J. Shieh and A. R. Hedges, J. Macromol. Sci. - Pure Appl. Chem., 1996, **A33**, 673.
33. J. Jane, Trends in Food Sci. and Technol., 1992, **3**, 145.
34. E. L. Powell, in Starch: Chemistry and Technology. Vol. II. 1st Ed. Ed. R. L. Whistler and E. F. Paschall. Academic Press, New York, NY, 1967, P. 523.
35. E. Pitchon, J.D. O'Rourke, and T.H. Joseph, U.S. Patent 4,280,851, 1981.
36. J. E. Eastman, and C.O. Moore, EP Patent 0,110,549 A2, 1984.
37. J. Jane, S. A. S. Craig, P. A. Seib, and R. C. Hoseney, Starch/Staerke, 1986, **38**, 258.
38. J. Chen and J. Jane, Cereal Chem., 1994, **71**, 618.
39. S. Rajagopalan & P. A. Seib, J. Food Sci., 1992, **16**, 13.

40. J. Chen and J. Jane, Cereal Chem., 1994, **71**, 623.
41. K. J. Zeleznak and R. C. Hoseney, Cereal Chem., 1987, **64**, 121.
42. H. Levine and L. Slade, Carbohydr. Polym., 1986, **6**, 213.
43. Y.-J. Wang and J. Jane, Cereal Chem., 1994, **71**, 527.
44. N. Inouchi, D. V. Glover, Y. Sugimoto, and H. Fuwa, Starch/Staerke, 1991, **43**, 468.
45. Y.-J. Wang, P. J. White, L. Pollak, and J. Jane, Cereal Chem., 1993, **70**, 171.
46. R. L. Shogren & B. K. Jasberg, J. Environ. Polym. Degrad., 1994, **2**, 99.
47. J. Jane, L. Shen, J. Chen, S. Lim, T. Kasemsuwan, and W. K. Nip, Cereal Chem., 1992, **69**, 528.
48. B. Pfannemuller, H. Mayerhofer, and R. C. Schulz, Biopolymers, 1971, **10**, 243.
49. M. J. Gidley and P. V. Bulpin, Carbohydr. Res., 1987, **161**, 291.
50. K. E. Spence, J. Jane, and A. L. Pometto III, J. Environm. Polym. Degrad., 1995, **3**, 69.
51. H. Bizot, B. Leroux, P. Le Bail, and A. Buleon, in Starch Structure and Function Conference Proceedings. Cambridge, UK, 1996.
52. H. N. Englyst and J. H. Cummings, Am. J. Clin. Nutr., 1987, **45**, 423.
53. D. Sievert and Y. Pomeranz, Cereal Chem., 1989, **66**, 342.
54. C.-Y. Lii and S.-M. Chang, J. Food Sci., 1981, **46**, 79.

Journal Paper No. J-16877 of the Iowa Agriculture and Home Economics Experiment Station, Amse, Iowa. Project No. 3258, and supported by Hatch Act and State of Iowa funds.

FUNCTIONAL PROPERTIES OF CATIONIC PEA STARCH

F.W. Sosulski, C. Yook and G.C. Arganosa

Department of Crop Science and Plant Ecology
University of Saskatchewan
51 Campus Drive
Saskatoon, SK Canada S7N 5A8

1　PRODUCTION OF FIELD PEAS

During the post-war era, agriculture in Western Canada became almost totally dependent on the production and export of wheat to Europe. As legislation and new baking technology restricted this market for strong gluten wheats, research scientists have developed alternate crops such as two-row barley and canola (Table 1). Whereas flour milling had almost disappeared as a local industry, malting barley and oil extraction plants are processing a high proportion of these new crops. The third and most recent diversification in crop production has been with starchy legumes or pulses, especially field peas and lentils. Markets for these seeds are in Asia, the Middle East and southern Europe, but options for value-added processing have been very limited. There are a few dehulling and pea splitting plants, and there is one protein/starch air classification facility and one wet milling operation, but none of these industries are thriving or expanding. The fiber and protein fractions from field peas have well established nutritional and functional properties but legume starches in general lack the desired functionality that are provided by commercial starches such as tapioca, potato, corn and wheat.

2　POTENTIAL UTILIZATION OF PEAS

The first limitation on utilization of pea starch is related to starch content in the seed (Table 2). Note that roots and tubers provide over 80% starch during extraction. Peas and other legumes contain less than 50% starch, so other components like protein and dietary fiber must be highly value-added for wet milling to be an economic proposition. Barley is in a similar position: low starch content and questionable commercial value of other grain components.

The physical characteristics of pea starch are favorable. The granules are uniformly large, with smooth surfaces and an oval shape. These dense granules exhibit no tendency to gelatinize and so a range of extraction conditions can be employed. In wheat and barley, a significant proportion of the granules are very small, yielding B-starch fractions that are of lesser value than the A-starch. Only corn granules have a similar uniform granule size to that of legumes but are much smaller in granule size.

Table 1 *1995 Crop Production in Saskatchewan*

Crop	Million Tonnes	Crop	Million Tonnes
Wheat	12.6	Canola	2.8
Barley	4.3	Flax	0.7
Oats	1.0	Mustard	0.1
Rye	0.2	Field Pea	0.9
Canary	0.1	Lentil	0.4

Table 2 *Chemical Composition of Starch Sources, % db*

Species	Organ	Starch	Protein	Fiber
Cassava	Root	88	4	4
Potato	Tuber	83	9	4
Corn	Grain	71	10	7
Wheat	Grain	62	16	12
Barley	Grain	55	15	20
Pea	Seed	46	23	16

In cross section, the granules exhibit a large vacuole, and a highly ordered crystalline structure has been demonstrated by x-ray diffraction[1]. Chemically, the outstanding feature of legume starches is their high amylose content, averaging about 35% (Table 3). These levels are substantially higher than is found in cereals (23-28%) and roots and tubers (<20%) but less than in amylocorn (70-80%). The degrees of polymerization of amylose in pea starches average 1300 units that is intermediate between corn and wheat but much below that of root and tuber starches. Legume starches have essentially no lipid or mineral components except on granule surfaces.

In North America, corn starch represents 95% of total starch production, exceeding 2.5M tonnes annually, and sets the price standard for competing starches like wheat (Table 3). Tapioca and potato starches are imported primarily and that is reflected in the Chicago prices quoted here for January, 1996. The corn utilization pattern in North America indicates the potential applications for any new starch source (Table 4). Legume starches including field pea are less digestible than tapioca or cereal starches when hydrolysed with pancreatic α-amylase[1,2]. During a 6-hr digestion period, corn starch was hydrolysed to the extent of 75% compared to 26-35% for legume starches. Thus, applications as sweeteners or dextrins are not an economic option for pea starch.

In food and industrial applications, the ease of granule dispersibility is an important prerequisite. The highly ordered crystalline structure of the granule limits the rate of water uptake and swelling power of pea starch[1,2]. Pea granules are slow to disperse, and solubility of the amylose and amylopectin is incomplete when heated to 95°C. (About 120°C will solubilize pea starch.) When heated at 95°C for 30 min, the cooked pastes are a mixture of intact granules, granule fragments and molecular starch that have no strong functional properties.

Table 3 *Properties of Commercial Starches*

Starch Species	Diameter µm	Amylose %	Amylose DP	P %	Price $U.S./tonne
Tapioca	25	17	3,000	.01	700
Potato	40	21	3,200	.08	750
Corn	15	28	940	.02	520
Waxy corn	15	2	-----	.01	-----
Wheat	25	26	2,100	.06	520
Barley	23	26	1,850	.02	430
Pea	30	35	1,300	.02	-----

In summary, legume starches have few commercial applications because of their slow water uptake, restricted swelling power, poor granule dispersibility, low molecular solubility, high gelatinization temperatures, stable amylograph viscosities, tendency to syneresis during storage, and resistance to α-amylase attack.

3   CHEMICAL MODIFICATION OF PEA STARCH

Functional properties of starches are often modified by covalently bonding organic and inorganic chemicals onto starch hydroxyl sites. Acetylation[3], hydroxypropylation[4], phosphorylation and cross-linking[5], and complexation with palmitate and glyceryl monopalmitate[6] resulted in limited improvement in thermal properties of legume starches.

A recent survey identified a current market for 40,000 tonnes per year of starch for the pulp and paper industry in Western Canada. Cationic starches are widely used as wet-end additives in paper making to improve sheet strength. A preliminary study of cationization of pea starch by Yook *et al.*[7] demonstrated that the positive charge on the free hydroxyl sites improved the water uptake and thermal properties to a marked degree.

Cationic starches are prepared by etherification with a tertiary or quaternary ammonium reagent[8]. The positively charged reagent reacts with the free hydroxy groups of starch chains in the granule under alkaline conditions (Scheme 1). The degrees of substitution (DS) are typically about 0.03, that is three cationic groups per 100 glucose units in starch.

Cationization at DS levels as low as 0.02 reduced the pasting and gelatinization temperatures, increased peak viscosity and set-back on cooling (Table 5). This phenomena was enhanced further by increasing DS to 0.04 and 0.06, except for lower set-backs due to interference with gel formation by the derivative.

The principal effects of cationization were to promote rapid granule dispersion at low pasting temperatures, yielding a molecular dispersion of amylose and amylopectin on heating to 95°C. On cooling, the gel structures were firm and, on storage for 7 days at 4°C and -15 °C, controlled syneresis (Table 6).

Table 4 *Uses of Corn Starch in United States*

Percentage	Use	Example
65%	Sweeteners	corn syrup (HFCS), dextrose sugar
5%	Food starch	bakery, dairy
1%	Adhesives	dextrins
30%	Industrial starch	paper, chemicals
	Fuel ethanol	

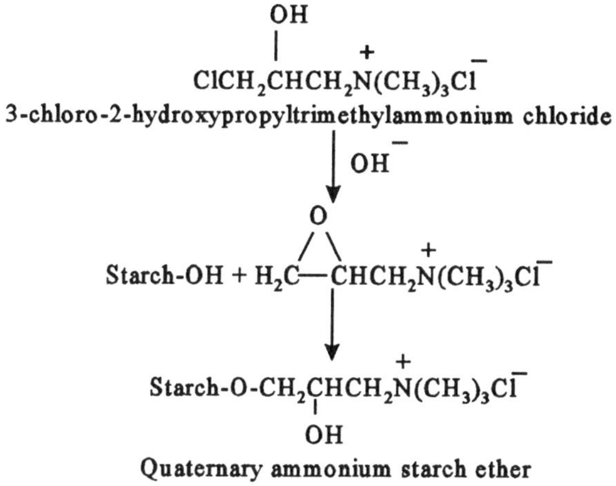

$$\text{OH}$$
$$|$$
$$\text{ClCH}_2\text{CHCH}_2\overset{+}{\text{N}}(\text{CH}_3)_3\overset{-}{\text{Cl}}$$

3-chloro-2-hydroxypropyltrimethylammonium chloride

$$\downarrow \quad \overset{-}{\text{OH}}$$

$$\text{O}$$
$$/ \ \quad +$$
$$\text{Starch-OH} + \text{H}_2\text{C}\!-\!\text{CHCH}_2\overset{+}{\text{N}}(\text{CH}_3)_3\overset{-}{\text{Cl}}$$

$$\downarrow$$

$$\text{Starch-O-CH}_2\overset{+}{\text{CHCH}_2}\text{N}(\text{CH}_3)_3\overset{-}{\text{Cl}}$$
$$|$$
$$\text{OH}$$

Quaternary ammonium starch ether

Scheme 1

Table 5 *Pasting Characteristics of Native and Cationic Corn and Pea Starches in 8% Aqueous Suspensions at pH 5.5*

Type and DS of Starch	Pasting Temperature °C	Peak Viscosity BU[1]	Viscosity at 95°C BU	Viscosity after 30 min at 95°C BU	Viscosity at 50°C BU
Corn-native	75	675	652	561	1290
Pea-native	71	-----	206	314	604
Pea-0.02[2]	59	2000	1478	1132	1835
Pea-0.04	50	2112	1140	730	1110
Pea-0.06	43	2280	852	610	924

[1] Brabender Units. [2] Number of cationic groups/glucose unit.

Table 6 *Degrees of Syneresis of Starch Gels after Storage*

Degrees of	7 days at 4°C		7 days at -15°C	
Substitution	Corn	Pea	Corn	Pea
0.00	13.4	24.3	8.3	16.0
0.02	0.3	0.0	0.0	0.4
0.04	0.0	0.0	0.0	0.3
0.06	0.0	0.0	0.0	0.0

4 FUNCTIONALITY OF CATIONIC PEA STARCH

Cationic starches are important derivatives used by the paper industry as wet end additives, surface sizes and coating binders. As an internal binder, the positive charges neutralize negative charges on pulp fibers, fillers and chemical additives. Thus, hydrogen bonding between fibers, and with starch as an adhesive, is achieved more efficiently than when no starch is added. Amphoteric starches containing both positive and negative charges are usually more effective than cationic starches for many functions in paper manufacture.

Cationic pea starch was evaluated as an internal binder in paper-making by the Paper Science and Engineering Department, Miami University, Oxford, OH. Cato 15, a commercial cationized (amphoteric) corn starch was used as a comparative treatment, along with a no-starch control and a cationic corn starch prepared under the same conditions as pea starch.

The cationic starches (DS=0.03) were added at 0.91% (20lb/ton^{-1} dry fiber) of fiber and five of the measures of paper strength are reported in Table 7. Cato 15 improved each of the measures of paper strength, compared to the no-starch control. Our laboratory-prepared cationic corn starch gave even better paper strength than Cato 15. However, cationic pea starch gave the highest values in three of the five parameters, and appeared to be a superior starch for paper manufacture.

5 CONCLUSIONS

The benefits of cationization were to promote rapid water uptake, granule swelling, dispersion and solubility of pea starch. The cooked pastes solidified into clear gels that showed high stability during storage, with essentially no evidence of syneresis.

Table 7 *Effect of 1% Cationic Starch on Paper Sheet Strength*

Cationic Starch DS=0.03	Breaking Length (km)	Tensile Index N*m/g	Burst Index kPa*m^2/g	Scott Bond ft*lb	TEA Yield Stress (psi)
No Starch	5.8	57	2.4	109	4590
Cato 15 (P)	6.7	66	3.1	177	4760
Corn	7.0	69	3.2	193	5370
Pea	7.6	74	3.1	170	6000

Cationic pea starch improved paper strength more than cationic or amphoteric corn starch. It should give better retention of fines, recycled paper and fillers. Also the drainage of fiber flocks, as they are processed into paper, would be enhanced. The overall effect is to reduce waste, effluent and pollution.

6   REFERENCES

1. R. Hoover and F. Sosulski, *Starch/Stärke*, 1985a, **37**, 181.
2. R. Hoover and F.W. Sosulski, *Can. J. Physiol. Pharmacol.*, 1991, **69**, 79.
3. R. Hoover and F. Sosulski, *Starch/Stärke*, 1985b, **37**, 397.
4. R. Hoover, D. Hannouz and F.W. Sosulski, *Starch/Stärke*, 1988, **40**, 383.
5. R. Hoover and F. Sosulski, *Starch/Stärke*, 1986, **38**, 149.
6. F. Sosulski, W. Waczkowski and R. Hoover, *Starch/Stärke*, 1989, **41**, 135.
7. C. Yook, F. Sosulski and P.R. Bhirud, *Starch/Stärke*, 1994, **46**, 393.
8. H.E. Carr and M.O. Bagby, *Starch/Stärke*, 1981, **33**, 310.

FUNCTIONAL AND PHYSICO-CHEMICAL PROPERTIES OF SOUR CASSAVA STARCH

C. Mestres[1], N. Zakhia[2] and D. Dufour[2, 3]

[1]CIRAD-CA, Laboratoire de Technologie des Céréales, Maison de la Technologie, BP 5035, 34032 Montpellier Cedex 1, FRANCE

[2]CIRAD-SAR, Unité de Recherche Technologie et Procédés, 73 rue J.F. Breton, 34032 Montpellier Cedex 1, FRANCE

[3]CIAT, Utilizacion de la Yuca, Apartado Aereo 6713, Cali, COLOMBIE

1. INTRODUCTION

Sour cassava starch, a traditional product of South America, is obtained after natural lactic fermentation and sun-drying. When a dough made with sour starch, mixed with green cheese and other minor components (milk, eggs), is oven-baked, an expansion occurs giving to the product an alveolar crumb structure. Typical bread-like products are thus daily prepared in Colombia and Brazil. Actually, only fermented and sun-dried cassava starch has this particular baking expansion ability. This study was undertaken in order to comprehend how functional properties of cassava starch are modified during natural fermentation and drying, and how this should be responsible for the unique baking expansion ability of sour starch.

2. MATERIALS

Cassava starch was extracted in a workers unit in Colombia. It was then portioned out in three samples that were submitted to various fermentation and drying conditions (Table 1).

Table 1 *Characteristics of cassava starches*

Sample	Fermentation step	Drying step	Baking expansion ability (cm³/g db)		
			pH 4 buffer	Water	pH 7 buffer
Sweet starch	None	Open air under cover of the sun (*ca* 30-35°C, 24 h)	5.3	6.3	3.5
Fermented/oven-dried starch	*ca* 20°C, 15 days	Oven at 40°C (8 h)	5.7	7.0	3.5
Sour starch	*ca* 20°C, 15 days	Under the sun (*ca* 40°C, 8 h)	16.7	14.9	9.7

In order to evaluate the baking expansion ability of the different samples, a dough was made by mixing starch with water and hydroxy-propyl-methyl-cellulose, respectively

48.2, 51.1 and 0.7 %. The specific volume (cm^3/g) of small balls of dough (20 g wb) was then measured after oven-baking at 290°C for 27 min. The baking expansion ability of sour starch was twice higher than that of unfermented (sweet) and fermented/oven-dried starches (Table 1); these two last samples had very similar baking expansion ability. This confirmed that fermentation and sun-drying are necessary to confer baking expansion ability to cassava starch[1, 2]. This property was further improved when the dough was acidified (pH 4), but made worst when the dough was neutralized.

3. HOT PASTE BEHAVIOUR

Starch slurries (7.1 % dry matter) were prepared in pH 4 and pH 7 buffers, and the evolution of their apparent viscosity was recorded using a Rapid Visco Analyzer (RVA, Newport Scientific, Narrabeen, Australia) with the following profile: holding at 35°C for 3 min, heating to 95°C at 6°C min^{-1}, holding at 95°C for 5 min, then cooling to 50°C at 6°C min^{-1} (Figure 1).

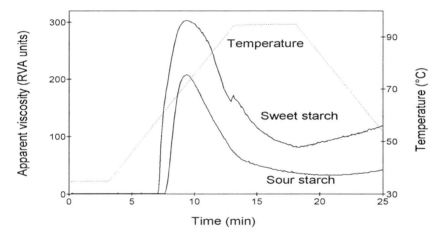

Figure 1 *Hot paste viscosity profiles of 7.1 % d. m. cassava starch slurries (in pH 4)*

Accordingly with previous results[1, 2], the apparent viscosity of sour starch paste was lower than that of sweet or fermented/oven-dried starch pastes. For example in acidic pH, the apparent viscosity at the beginning of the plateau at 95°C was twice lower for sour starch (Table 2). Also, sour starch paste did not present any set back phenomenon during cooling, on contrary to sweet starch (Figure 1). Furthermore, these differences were enhanced at pH 7: the viscosity of sweet and fermented/oven-dried starch pastes increased slightly in neutral pH whereas viscosity of sour starch paste was dramatically lowered when it was prepared with pH 7 buffer (Table 2).

Table 2. *Apparent viscosity of 7.1 % d.m. cassava starch pastes measured at the beginning of the plateau at 95°C*

Sample	Apparent viscosity (RVA units)	
	in pH 4 buffer	in pH 7 buffer
Sweet starch	167	175
Fermented/oven-dried starch	147	157
Sour starch	77	32

3.5 % dry matter slurries were pasted at 65°C in pH 4 and pH 7 buffers for measuring starch swelling and solubility (Table 3). Solubilized starch concentration and amylose ratio in the solvent phase were determined after iodine complexation and absorbance readings at 545 and 620 nm.

Table 3 *Cassava starch swelling and solubility values after pasting at 65°C*

Sample	in pH 4 buffer			in pH 7 buffer		
	Swelling power in the dispersed phase (g/g)	Starch concentration in solvent phase (mg/mL)	Amylose percentage in solvent phase (% starch basis)	Swelling power in the dispersed phase (g/g)	Starch concentration in solvent phase (mg/mL)	Amylose percentage in solvent phase (% starch basis)
Sweet starch	22	3.6	91	16	3.5	78
Fermented/oven-dried starch	18	3.6	70	14	3.5	70
Sour starch	18	6.8	48	17	12	41

No clear differences were shown between starch swelling indexes of the three samples. However, starch solubilisation was two to three times higher for sour starch paste than for sweet starch and fermented/oven-dried starch pastes; this effect was higher in neutral pH. Soluble starch was mainly constituted of amylose for sweet starch paste (leaching phenomenon) whereas it was, at least, half amylopectin for sour starch paste. Hence, the higher starch solubilisation observed for sour starch paste, comparing to that of sweet starch, was mainly due to higher amylopectin solubilisation. This could explain the lower hot paste viscosity of sour starch paste; indeed, amylopectin solubilisation should induce a softening of swollen starch ghosts and hence softening of the paste. It was confirmed, with additional cassava starch samples (obtained after fermentation and drying the sweet cassava starch with various conditions), that hot paste viscosity dramatically decreases when amylopectin solubilisation increases (Figure 2).

Figure 2 *Relationship between the amylopectin concentration in solvent phase and apparent viscosity of cassava starch pastes measured at the beginning of the plateau at 95°C*

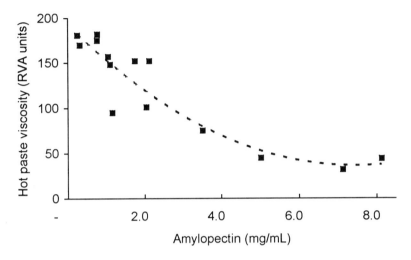

4. MACROMOLECULAR CHARACTERIZATION OF CASSAVA STARCH

4.1. Intrinsic viscosity

Intrinsic viscosity of cassava starch samples was determined at 35°C in 0.2 M potassium hydroxyde by extrapolation to zero concentration of the reduced and inherent viscosities. Samples were first solubilised in de-oxygenated 1.0 M potassium hydroxide under constant agitation for 6 days then diluted five times with ultrapure water a day before testing. Samples were then filtered through 5 μm pore size filter. Accordingly with previous results[2], sour starch intrinsic viscosity appeared almost twice lower than that of sweet starch (Table 4) or than that of fermented/oven-dried sample (207 mL/g).

Table 4 *Intrinsic viscosity of cassava starches*

Sample	Intrinsic viscosity (mL/g)		
	Raw	Purified	Amorphous
Sweet starch	198	239	237
Sour starch	131	125	128

Sweet and sour starches were purified by successive washings with various solvents then sieved through 125 μm sieve (Figure 3). Non-granular and amorphous starches were obtained by solubilization with dimethyl-sulfoxide followed by ethanol precipitation. The intrinsic viscosity of sweet starch slightly increased after the purification procedure. However, intrinsic viscosity of sour starch was kept constant after the purification and destructuration procedures, and remained almost twice lower than that of sweet starch. This indicated that the lower intrinsic viscosity of sour starch was not linked to some interactions with minor components that were leached out by the washing procedure.

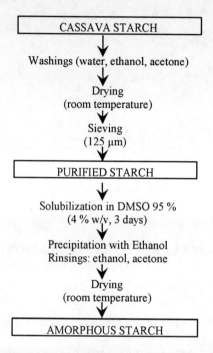

Figure 3 *Purification and destructuration procedure of cassava starches*

Amylopectin and amylose were then extracted and purified from amorphous sweet and sour starches according to the procedure of Banks and Greenwood[3] (Figure 4). Their intrinsic viscosities were determined (Table 5) in the same conditions as for starch samples. Intrinsic viscosities of both macromolecules extracted from sour starch were lower than those extracted from sweet starch. In particular, the intrinsic viscosity of sour starch amylose was almost three times lower than that of sweet starch.

Table 5 *Intrinsic viscosity (mL/g) of amylose and amylopectin extracted from cassava starches*

Sample	Amylopectin	Amylose
Sweet starch	176	496
Sour starch	121	177

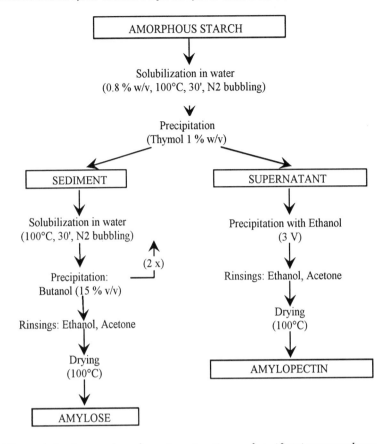

Figure 4 *Amylose and amylopectin extraction and purification provedures*

Incidentally, it should be noted that both purified amylopectins did inflate during final drying at 100°C whereas purified amylose did not. This clearly indicated that gas retention ability is linked to the amylopectin fraction of starch.

4.2. Gel permeation chromatography

Gel permeation chromatography (GPC) of amylose and amylopectin were performed using Fractogel HW-65 (S) and HW-75 (S) respectively. Samples were solubilised as previously described for intrinsic viscosimetry measurements. Elution was achieved at 35°C with 0.2 M potassium hydroxide at 0.6 mL min[-1] using refractometric detection[2].

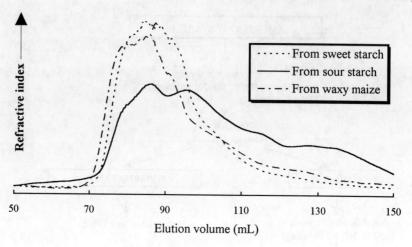

Figure 5 *Comparison of GPC elution patterns of amylopectins extracted from cassava and waxy maize starches*

Sweet starch amylopectin gave one broad peak (Figure 5) that began at the void volume of the column (close to 70 mL). This profile was very similar to that observed for waxy maize amylopectin. The elution pattern was quite different for sour starch amylopectin, which presented multiple peaks (at 88, 98, 115 and 130 mL) profile. It is likely that a large part of amylopectin has been slightly altered due to the joint action of fermentation and drying stages. However, no peak was detected at the total volume (150 mL), indicating that no small molecules were present.

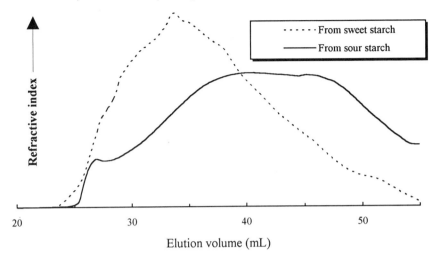

Figure 6 *Comparison of* GPC *elution patterns of amyloses extracted from cassava starches*

A very similar conclusion could be drawn for amylose. GPC elution pattern of sweet amylose presented one broad peak (Figure 6) with a maximum close to 35 mL. For sour starch amylose, a multiple peak elution pattern was observed. The first peak, at 28 mL, was likely due to amylopectin contamination. Then a plateau between 40 and 48 mL was observed whereas no peak at total volume (55 mL) could be evidenced. This suggests, as for amylopectin, that a large part of amylose has been altered during processing.

5. CONCLUSION

Functional properties of cassava starch are dramatically changed by the joint action of fermentation and sun-drying processes. In particular, sour starch amylopectin is much more soluble than that of sweet starch. This induces a softening of swollen starch ghosts during pasting and hence softening of sour starch hot paste.

The macromolecular structure and/or conformation of amylose and amylopectin are also dramatically changed by cassava starch fermentation and sun-drying. Their intrinsic viscosity decreases, particularly for amylose (three times lower than that of sweet starch). Gel permeation chromatography shows that this drop was not due to an extensive hydrolysis of one part of the macromolecules, but on contrary that the whole macromolecule population is affected.

These changes in functional properties and macromolecular structure of cassava starch may be due to a partial hydrolysis (by a photo-oxidate phenomenon for example) of the macromolecules during sun-drying. However, this hypothesis cannot explain the crop of sour starch hot paste viscosity when the pH increases from 4 to 7.

Purified cassava amylopectin has a baking expansion ability. This allows us to formulate a hypothesis explaining the gas retention ability of sour cassava starch. The joint action of fermentation and sun-drying could increase amylopectin solubilisation ability (due to its partial hydrolysis for example). The solubilised amylopectin could then act as a stabiliser for the gas bubbles formed during cooking. Further studies will be developed in order to validate this theory.

References

1.   D. Dufour, C. Brabet, N. Zakhia and G. Chuzel, in 'Transformation alimentaire du manioc', T. Agbor Egbe, A. Brauman, D. Griffon and S. Treche eds, ORSTOM, Paris, 1995, p 399.
2.   C. Mestres and X. Rouau, *J. Sci. Food Sci.,* 1996 (in press)
3.   W. Banks and C.T. Greenwood, *Starch* , 1967, **19**, 394.

STRUCTURAL AND POLYMORPHIC TRANSITIONS OF AMYLOSE INDUCED BY WATER AND TEMPERATURE CHANGES

P. Le Bail, A. Buleon, P. Colonna and H. Bizot

Institut de la Recherche Agronomique, BP 1627-44316 Nantes, France

1 INTRODUCTION

The polymorphism of starch, as revealed by X-ray diffraction analysis, has been recognized for a long time and three main forms A, B and V have been reported[1-4]. The A polymorph occurs frequently in cereal starches, while the B polymorph is characteristic of tuber and amylose-rich starches. The V polymorph essentially results from the complexation of amylose with compounds such as iodine, alcohols or lipids, and is rarely detected as a crystalline material in native starches[5,6].

Amylose can be recrystallized from solution in A, B or V form[7]. This ability has been used to determine the three dimensional structures of these different polymorphs, and to investigate the structure of crystalline domains in native starches. In A and B forms, a double helical conformation has been proposed for amylose chains by Wu & Sarko[8,9]. More recently, new modified packing models were proposed by Imberty et al.[10] for amylose A and Imberty & Perez[11] for amylose B. In the A-type structure, left handed parallel-stranded double helices are packed in the monoclinic space group B2. In the B-type structure, the double helices are packed in a hexagonal unit cell with the $P6_1$ space group.

Some of the V-type structures (Va, Vh, Viodine...) have been determined by X-ray analysis on crystalline fibres[12-14] or by electron crystallography[15]. The chain conformation consists in a left-handed six residues per turn helix with a rise per monomer between 0.132 and 0.136nm. The characteristic Vh structure obtained with linear alcohols and fatty acids has been extensively studied. In the currently accepted model[13], helices are packed in the orthorhombic space group $P2_12_12_1$.

In the solid state, some polymorphic transitions may occur between amylose polymorphs in specific conditions of hydration and temperature. B-type to A-type transitions have been described during heat moisture treatment on both native starches[16,17] and crystalline residues of native starches[18]. These transitions occur usually upon heating at low and intermediate water contents as various successive endothermic transitions including also recrystallizations and multiple melting behaviour[19-21]. On the contrary, starch annealing (ie : thermal treatment at high water contents at temperatures slightly below the gelatinization temperature) was shown to increase the melting temperature and to narrow the melting domain[22-25], but not to increase crystallinity or to modify the crystalline type[26].

The V-type structures could be very prone to polymorphic conversions induced by rehydration or drying without heating. The Vh to Va conversion is observed when drying the Vh form at a water activity lower than 0.6[13,27,28]. This transition was observed a long time ago on fibrillar crystals prepared from amylose-butanol complexes or from solution in DMSO[27,29], and more recently, on polycrystalline powders of amylose-alcohol complexes during storage at low water activity[30]. It is usually interpreted in terms of shrinkage of the unit cell upon removal of water molecules present between amylose helices. Transformation of the Vh-type into the B-type was also observed upon rehydration[29,30] but this conversion involving both double helical and single helical conformations is not interpreted.

This paper describes some polymorphic and phase transitions induced by hydrothermal treatment (25-180°C / 10-45 %H$_2$O) or water activity changes involving A-, B- and V-type structures. Most results were obtained on amylose fractions with different degrees of polymerization (DP), and a few of them also on native starches.

2 - B TO A TRANSITIONS

A polymorphic conversion from the B- to the A-type[18] can be produced by heating (heating rate 3°C.min^{-1}) lintnerized potato starch (average degree of polymerization (DP) : 15, - B-type) at moisture contents between 20 and 45 %H$_2$O . Using sequential heat treatments in Differential Scanning Calorimetry (D.S.C.) and a characterization of the corresponding heated products by X-ray diffraction, the conversion was shown to result in a small endotherm with enthalpy always lower than 4J/g[18] at temperatures varying from 88°C to 105°C depending on the water content. Figure 1a shows the X-ray diffraction data obtained at 42 %H$_2$O. The transition occurred at 88.6 °C and the corresponding enthalpy was 4 J/g. The X-ray diffraction diagrams of the sample heated at 107°C and 130°C were characteristic of a mixture of A- and B-types and pure A-type respectively.

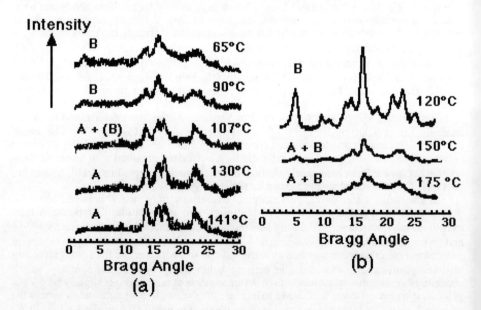

Figure 1 *X-ray diffraction diagrams showing the B- to A-type transition : a) lintnerized potato starch, b) lintnerized wrinkled pea starch.*

At similar water contents, this conversion occurs at a higher temperature (145-160°C) for lintnerized wrinkled pea starch which has an average DP 30 and a B-type crystallinity[31]. But, in that case, the conversion is not complete before melting as shown in Figure 1b; the diffraction diagrams obtained on substrates heated at 150°C and 175°C respectively are characteristic of a mixture of A- and B-types while incipient melting was observed between 175 and 185°C.

A similar transition is induced in the same conditions of temperature and hydration for corresponding native starches, but the transition overlaps with melting for both potato (Figure 2) and wrinkled pea starches. This phenomenon is very similar to that which occurs during heat moisture treatments of native starches. This process is well known to modify the functional properties of starch and to change the structural type of B-type starches[32]. The hydration and temperature conditions involved (100-120°C, 18-24 %H_2O) are similar; nevertheless, the kinetics is strongly different since the treatment time required ranges usually between 18H and 24H.

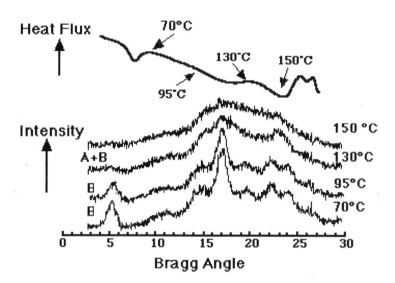

Figure 2 *D.S.C. thermogram of native potato starch at 22% H_2O (heating rate 3°C.min^{-1}) and corresponding X-ray diffraction diagrams showing the B- to C-type transition above 100°C.*

3 - ANNEALING AND RECRYSTALLIZATION

For water contents lower than 20%, no polymorphic transition is observable for B-type native and lintnerized starches. On the other hand, an important crystallinity increase (annealing) is observed for A-type lintnerized starches at 17% hydration between 100 and 120°C and, to a lesser extent, for lintnerized potato starch at 12 %H_2O. A similar annealing is also observed at intermediate water contents (20-45 %H_2O) after the B to A transition, when newly formed A-type structures are heated below melting temperature as shown in Figure 1.

Such a recrystallization has never been described on native starches whatever water content used. Below 20 %H_2O, no structural transition is obvious before melting. At intermediate water contents, the crystallinity of A-type starches decreases slightly upon heating or does not change and the polymorphic type of B starches is strongly modified as described above. In excess water, the crystallinity is kept intact or slightly decreases irrespective of the polymorphic type (Figure 3). Therefore, the changes in melting behaviour brought about by annealing in excess water (that is increase of the melting temperature and decrease of the melting domain) cannot be ascribed to recrystallization.

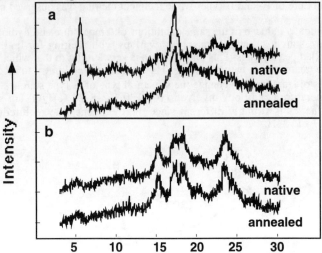

Figure 3 *X-ray diffraction diagrams of native and annealed (in excess water) potato (a) and wheat (b) starches.*

4 - A TO Vh TRANSITION

Heating native maize starch at intermediate moisture contents (between 19 and 30%) does not induce any improvement of crystallinity as mentioned above but a polymorphic conversion from the A- to the -Vh type as shown in Figure 4 for 19%H_2O. In this case, the

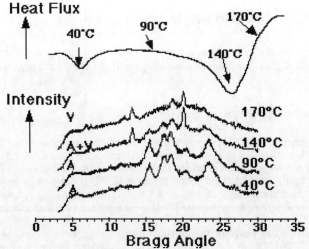

Figure 4 *D.S.C. thermogram of native maize starch at 25% H_2O (heating rate : 3°C.min^{-1}) and corresponding X-ray diffraction diagrams showing the A- to V-type transition above 90°C.*

appearance of the Vh structure is due to the complexation of amylose by lipids present in starch. At 19%H_2O, the main endotherm with temperature onset close to 130°C corresponds to the beginning of starch melting and to the crystallization of amylose-lipid complexes. Between 150 and 170°C, the amount and the crystallinity of complexed amylose increase and melting of the A-type structures occurs.

5 - POLYMORPHIC TRANSITIONS FROM Vh-TYPE

5.1. Vh to Va transition

This conversion is well known and was described for both fibrillar crystals[13,27,28,29] and polycrystalline powders[30] of Vh-type amylose complexes. It is induced by drying or storage at water activity lower than 0.6. The kinetics is strongly dependent on amylose chain length; a complete conversion was described[30] after vacuum drying during 24 hours for some complex prepared with DP900 amylose.

5.2. Vh to B transition

Vh-type amylose/alcohol and amylose/lipid complexes rearrange into the B-type upon conditioning at high water activity or by direct immersion in water. The amount of conversion and the kinetics depend on the chain length of amylose. In case of amylose-ethanol complex, it takes about 48 hours for DP 20 and 30 amylose while 7 days are needed for the total conversion of DP80 and DP900 amylose. An intermediate amorphous state is always observed for DP900. Such an intermediate decrystallization probably occurs in the case of DP80 but much too quickly to be observable. This polymorphic form can also be reached by direct immersion of the crystals into water for 48h except for the DP20 amylose which is almost dissolved[30].

Similar Vh to B transition can be obtained by storing amylose-ethanol complexes over mixtures of water-ethanol (Table 1). Nevertheless, in that case, Vh to A conversions are also

Table 1 *Evolution of the polymorphic type of Vh crystal upon storage over mixtures of water and ethanol.*

water ethanol ratio	10/90	30/70	50/50	90/10	100/0
initial type	type Vh	type Vh	type Vh	type Vh	type Vh
sample water and ethanol contents	13,86% water 42,12% ethanol	19,75% water 19,05% ethanol	24,48% water 11,68% ethanol	30,37% water 1,73% ethanol	33,3% water 0,9% ethanol
resulting type	type Vh	type Vh +(A)	type Vh + A	type B	type B

induced by higher concentrations of ethanol. When using 90/10 mixtures, the Vh form is stable, but for 30/70 and 50/50 ratios, diffraction diagrams are characteristic of a mixture of Vh and A-type, while a pure B-type is obtained when more than 90% water are present in the conditioning mixture. The corresponding contents in water and ethanol are presented in Table 1. The Vh form is stable when the relative ethanol content is higher than 75% and the A form appears when it was comprised between 50 and 33%. The B form resulted from strong rehydration.

For amylose-fatty acid complexes, the rate of conversion depends on both fatty acid and amylose chain lengths, increasing with decreasing chain lengths. This is shown for lauric acid complexes on Figure 5 where the conversion is complete for DP30 amylose while some traces of Vh-form are still detected in diffractograms of DP40-C12 complexes. A similar effect of the fatty acid chain length is observed with complete conversion when using caprylic acid (C8) and greater stability for lauric (C12) and palmitic (C16) acids. The combination of palmitic acid and DP 900 amylose yields stable Vh structures.

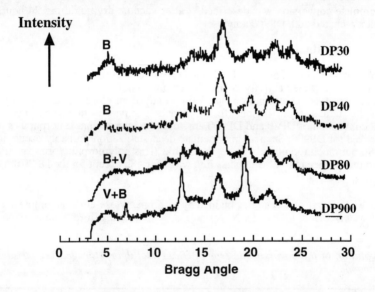

Figure 5 *X-ray diffraction diagrams of amylose fatty acid complexes prepared with different amylose chain lengths, showing the evolution from the Vh- to the B-type in presence of water.*

6 - CONCLUSION

In most cases, the transitions described above can be interpreted on the basis of the structural models established for starchy substrates. The B- to A-type conversion can be understood following two distinct mechanisms : 1) a dehydration of the B-type structure and subsequent reorganization of the crystalline packing into the more dense/less hydrated A-type and 2) a melting of the B-type organization and subsequent recrystallization of the melt into the A-type. The first hypothesis was proposed by Imberty et al[33], because of the presence of similar pairs of double helices in both A- and B-types. The c axis origins of the two double helices differ by about C/2, allowing a very close nesting of the crests and troughs of the paired double helices. In the transition, these pairs could be preserved and could rearrange into the A form. Thus, the weak endothermic event observed during the transition would correspond to the heat necessary for water removal and breakage of inter chain hydrogen bonding. The latter has been shown to be very low ($\cong 0.71 J/g$) for V-type amylose complexes[20].

The second mechanism is in agreement with the results of Whittam et al[34], on the melting behaviour of A- and B-type spherolitic crystals of amylose as a function of water content. The melting curves are parallel with a 20° C difference in melting temperatures of the two polymorphs, with the A-type always melting at higher temperature. In this case, the domain comprised between the two curves could correspond to the conditions for recrystallization of A-type structure from the melt. Nevertheless, referring to the low enthalpy associated to the B to A transition (between 0.2 and 4J/g[18]), which is far from the usual enthalpy measured for melting of crystalline starchy substrates (12-20J/G for native starches and up to 35 J/G for spherocrystals), only a high exothermic contribution of the recrystallization to the DSC signal is conceivable with such a mechanism. Finally, the more rapid conversion of lintnerized products could be interpreted in terms of greater mobility of short independent chains in comparison to the longer ones involved in the complex architecture and branched network of the native starch granule.

Recrystallization induced by hydrothermal treatments at temperatures below melting temperature Tm was only observed with lintnerized products which present shorter chain length and smaller cooperativity. This ability of independent chains to reorganize in the presence of heat and moisture is also encountered in the transition from free (isolated helices) to crystalline amylose complexes[20, 35]. In that case, it was proposed that the transition corresponds to the packing of separate complexed amylose chains into a more crystalline structure (propagation).

The crystalline rearrangements of native starches are more difficult because the chains are involved in the complex architecture of the starch granule and large moves are hampered by branching and steric hindrance. This is particularly true for native maize starch where no crystallinity increase is obvious before melting. But in that case, more interesting is the formation of the Vh type. This crystalline type is not observed on native starch before thermal treatment and crystallization is probably induced upon heating. The Vh structure formation could probably correspond to the small endotherm detected at 100°C and may occur following two mechanisms. The first one involves the presence in nature starch of separate complexed amylose chains as demonstrated by Morrison et al[6] for barley starch. Heating would provoke some propagation into crystalline packing of these isolated complexes as shown for amylose-fatty acid complexes prepared at lower temperatures. The second one refers to crystallization from amylose and lipid released during starch melting. In both cases, an endothermic phenomenon is observed, showing that partial melting is necessary to recrystallization.

The Vh to B and Vh to A transitions are more tricky to understand according to the present knowledge of starch structure, since they involve the change from a single to a double helical conformation. Such a mechanism is conceivable if the crystals are first dispersed or at least dissociated in the reaction medium. A rearrangement of free mobile chains similar to retrogradation could occur leading to the crystallization of the B type. An intermediate amorphous state was observed by Le Bail et al[30] on amylose-ethanol complexes prepared from high molecular weight amylose. Such a reorganization could occur more or less rapidly depending on the chain lengths involved and the solubility. A similar change could be invoked for the Vh to A transition which is observed for short chains of amylose which are probably less stable. For these materials, storage over water-ethanol mixtures is sufficient to induce the crystallization into A, B or Vh type depending on the respective concentrations of alcohol and water. In this case, the concentrations correspond to those used for crystallization of lamellar crystals of amylose[7,15]. In any case, the numerous structural transitions of starch including iodine complexing, make the double helical conformation proposed usually for A- and B-type structures questionable. For example conce ning the Vh to B-type transition, it is unlikely that any drastic conformational change such as defolding/refolding could be provoked by such mild treatment as rehydration, unless the helical conformations ascribed to Vh- and B-types are close to each other as recently proposed by Saito et al[36].

REFERENCES

1. J.R. Katz and T.B Van Itallie, *Z. Phyik. Chem.*, 1930, **A150**, 37.
2. D. French and V. Murphy , *Cereal Foods World*, 1977, **22**, 61.
3. F. Duprat, D. Galland, A. Guilbot, C. Mercier and J.P. Robin, In: "Les polymères végétaux". Ed. B. Monties, Gauthier-Villars, 1980.
4. A. Sarko and P. Zugenmaier, In: "Fiber diffraction methods". Eds A.D. French & K.H. Gardner, ACS Symposium series, Washington 1980.
5. C. Gernat, S. Radosta, H. Anger and G. Damashun, *Starch/Stärke*, 1993, **9**, 309.
6. W. Morrison, R. Law and C. Snape C., *J. Cereal Sci.*, 1993, **18**, 107.
7. A. Buleon, F. Duprat, F.P. Booy and H. Chanzy, *Carbohydr. Polym.*, 1984, **4**, 161.
8. H.C. Wu and A. Sarko A., *Carbohydr. Res.*, 1978, **61**, 27.
9. H.C. Wu and A. Sarko A., *Carbohydr. Res.*, 1978, **61**,7.
10. A. Imberty, H. Chanzy, S. Perez, A. Buleon and V. Tran, *J. Mol. Biol.*, 1988, **201**, 365.
11. A. Imberty and S. Perez, *Biopolymers*, 1988, **27**,1205.
12. W. Winter and A. Sarko, *Biopolymers*, 1974, **13**, 1447.
13. G. Rappenecker and P. Zugenmaier, *Carbohydr. Res.*, 1981, **89**,11.
14. T.L. Bluhm and P. Zugenmaier, *Carbohydr. Res.*, 1981, **89**, 1.
15. J. Brisson, H. Chanzy and W. Winter, *Int. J. Biol. Macromol.*,1991, **13**, 31.
16. L. Sair in *"Methods of carbohydrate chemistry"*, ed. Whistler R.L., Academic Press, New-York, 1964.
17. K. Kulp and K. Lorenz, *Cereal Chem.*, 1981, **58**, 46.
18. P. Le Bail, H. Bizot, A. Buleon, *Carbohydr. Polym.*, 1993, **21**, 99.
19. J. Donovan, *Biopolymers*, 1979, **18**, 263.
20. C.G. Biliaderis, C.M. Page, T.J. Maurice and B.O. Juliano, *J. Agric. Food Chem.*, 1986, **34**, 6.
21. P.L. Russel, *J. Cereal Sci.*, 1987, **6**, 133.
22. J.L. Marchand and J.M.V. Blanshard, *Starch/Stärke*, 1978, **30**, 257.
23. B.R. Krueger, C.A. Knutson, G.E. Inglett and C.E. Walker, *J. Food Sci.*, 1987, **52**, 715.
24. C.A. Knutson, *Cereal Chem.*, 1990, **67**, 376.
25. H. Jacobs, R. Eerlingen, W. Clauwaert and J. Delcour, *Cereal Chem.*, 1995, **72**, 480.
26. R. Stute, *Starch/Stärke*, 1992, **44**, (6), 205.
27. M.E.Hinkle and H. Zobel, *Biopolymers*, 1968, **6**, 1119.
28. V.G. Murphy, B. Zaslow and A.D. French, *Biopolymers*, 1975, **14**, 1487.
29. B. Zaslow, V.G. Murphy and A. French, *Biopolymers*, 1974, **13**, 779.
30. P. Le Bail, H. Bizot, B. Pontoire and A. Buleon, *Starch/Stärke*, 1995, **47**, 229.
31. P. Colonna, A. Buleon, C. Mercier and M. Lemaguer, *Carbohydr. polym.*, 1982, **2**, 43.
32. A. Kawabata, N. Takase, E. Miyoshi, S. Sawayama, T. Kimura and K. Kudo, *Starch/Stärke*, 1994, **46**, 463.
33. A. Imberty, A. Buléon, V. Tran, and S. Perez, *Starch/Stärke*, 1991, **43**, 375.
34. M. Whittam, T.R. Noel and S.G. Ring, *Int. J. Biol. Macromol.*, 1990, **12**, 359.
35. C.G. Biliaderis, *Food technol.*, 1992, **6**, 98.
36. H. Saito, J. Yamada, T. Yukumoto, H. Yajima and R. Endo, *Bull. Chem. Soc. Jpn*, 1991, **64**, 3528.

MOLECULAR CHANGE OF STARCH GRANULE WITH PHYSICAL TREATMENT. POTATO STARCH BY BALL-MILL TREATMENT

T. Yamada, S. Tamaki, M. Hisamatsu and K. Teranishi

Department of Food Chemistry, Faculty of Bioresources
Mie University
Tsu, 514, Japan

1 INTRODUCTION

Ball-mill treatment of starch granule was first investigated by Meuser[1] . He found that the starch granule was changed to plastic-like with compression, and he tried to change the compression system to ball-mill system. As the result, he showed that the treated starch granule became very sensitive to amylase and increased its solubility and these phenomena were resembled to gelatinization although he adopted vibration type ball-mill, hence large amount of damaged starch was produced. Therefore, few researchers noticed the importance of this report. At the start of our study, we also intend to make damaged starch model by ball-mill, and we noticed that this rolling type ball-mill treatment did not produce damaged starch but another one. And we knew Meuser's report. Rolling type ball-mill is far mild treatment than vibration type. Hence, it needs far more time to obtain same extent change on starch granule in the case of rolling type to vibration type. But, this type can produce relatively homogeneous one without damaged (so call) one. We started investigation of the mechanism. Almost same time, Morrison[2,3,4] noticed this phenomenon when he intended to make damaged wheat starch by ball-mill treatment. And he intensively investigated and made a series of reports concerning to this phenomenon comparing with damaged starch. His experimental result of ball-milled wheat starch is almost same to ours, but some differences remain. The most important problem is that existence of resistant granule to this treatment. Namely, we could not find such a resistant one, and we assumed that discrepancy would happen with ball-mill condition. Because, we found that the progress of change of starch condition by ball-mill was not homogeneous when we did not take care to remove starch attached to corner of the mill through the processing. And also we counted the number of treated starch granules suspended in water, and we found that the number was decreased to about 1/1000 of the original. This indicated that practically no difference among starch granules. We tried ball mill treatment further to 320h and found that such a long treatment brought to thorough change of molecular structure of starch components.

Hence, we intended to apply this treatment to potato starch. Raw potato starch is well-known to have resistance to amylase. However, when it is gelatinized, it becomes sensitive to amylase. This observation suggests that the resistance mechanism should be asked to starch granule structure. In spite of great advance in molecular structure of starch components, situation of amylose in starch granule is still remained ambiguous. We considered the ball mill treatment would be effective tool to elucidate this problem.

2. MATERIALS AND METHODS

Potato starch
Domestic commercial potato starch

Ball-mill and treatment condition
Rolling type, diameter; 17cm, height; 15cm, ceramic ball; 1.5cm (1.2kg), starch; 100g, rotatory speed; 90rpm, temperature; 25-35°C

Water absorption
Starch sample (500mg) was placed in centrifuge tube having inner sieve covered with filter paper. After suspending in water, the suspension was kept for 1h and centrifuged with 6000rpm. Water absorption was estimated from weight increase.

Amylase susceptivity
Starch sample (250mg) was suspended in water (50ml) and glucoamylase solution (0.4ml) (16unit) (industrial grade, from Rhizopus species) was added. The suspension was gently stirred on rotatory shaker at 40°C for 24h. Total saccharide and reducing power of supernatant separated from reaction mxture were measured by phenol-sulfuric acid method and Somogy-Nelson's method,respectively.

Differencial Scanning Calorimetry
Starch sample (15mg) was placed in the capsule (70µl, Ag) and water (45µl) was added and the capsule was sealed. DSC was measured by DSC-100P-SSC-5200 system(Seiko-Electric Co.) under the condition of constant temperature elevation speed (2°C/min).

GPC
Starch sample solution for GPC was prepared by perchloric acid method. Starch sample(50mg) was placed in a tube and 40% perchloric acid (0.5ml) was added. The starch was solubilized with trituration at 0°C. after neutralization, and the clear solution was subjected to GPC(Toyopearl-HW 75 and -65, respectively, (2.5x100cm)(solvent; water, elution speed;1ml/min). Eluate was fractionated to 10ml, and total saccharide and reducing power (Park Johnson's method) and absorption of iodine color reaction were measured.

Debranching
Debranching of starch sample was done by isoamylase(purified from commercial enzyme, Hayashibara). The debranched sample was subjected to GPC (Toyopearl-HW 50) under the same condition mentioned above.

3. RESULTS AND DISCUSSION

SEM of ball-milled starch sample are shown in Fig.1. The surface of starch granule is changed to rough after 5h, and smoothness of surface perfectly disappeared after 10h and appearance was changed. Some fragmentation occurred and granule size became less after 80h. Granules were coagulated after 320h. Adler[5] observed that potato starch granule was first disrupted from central region of granule and inner component was burst out by ball-mill treatment. We also observed same but the extent of this damage was not same among granules. Because, the shape of granule has so many varieties, and especially ellipsoid type was most damaged. This observation is reasonable when the ball more hit the center region in order to its wide area of surface. And small round type granule still remain in almost intact appearance after 10h.

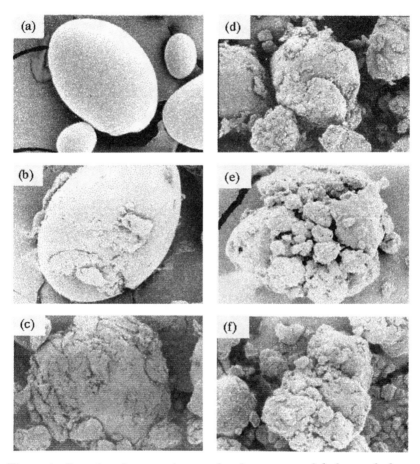

Figure 1 *Scanning electron micrographs of potato starch before and after*
ball-mill(B/M) treatment. (a) original starch ; (b) B/M 5h ;
(c) B/M 10h ; (d) B/M 20h ; (e) B/M 80h ; (f) B/M 320h.

X-ray diffraction patterns of samples are shown in Fig.2. And crystallinity of the
sample is shown in Table 1. Potato starch has B type pattern, and it retained the pattern
till 5h, but it almost loss after 10h and perfectly loss after 20h. Crystallinity of 10h
sample is 17.1% ,and that of 20h 12.8%, respecively.

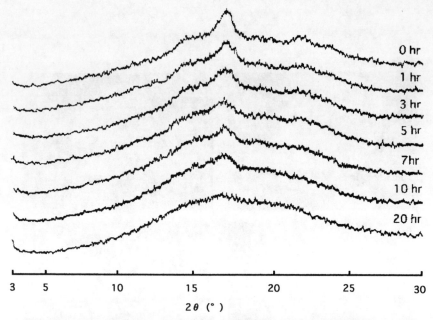

Figure 2 *X-ray Diffraction Patterns of Ballmill Treated Potato Starch*

Table 1 *Crystallinity of Ballmill Treated Potato Starch*

Potato (hr)	0	1	3	5	10	20
Crystallinity(%)	29.7	25.7	21.7	21.1	17.1	12.8

Water absorption ability is shown in Fig.3. This ability rapidly increased till 10h (200%) and slightly up at 20h (220%) and reached 280% after 80h.

Amylase sensitivity is shown in Fig.4. The tendency of production of total saccharide and reducing power is almost same. Ball-mill treatment time was the more, the more hydrolysis was observed. In any case of amylase treatment, hydrolysis rate became plateau over 7h amylolysis. Hence, ball-mill treatment does not uniformly affect every granule and intact part still remains. But, the sample of 80h was perfectly hydrolysed.

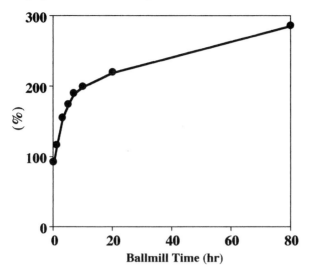

Figure 3 *Water Absorption of Ballmill Treated Potato Starch*

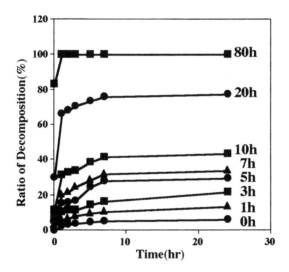

Figure 4 *Amylase Susceptibility (Glucoamylase)*
of the Ballmill Treated Potato Starch

DSC patterns and results of sample are shown in Fig.5 and Table 2, respectively. These results showed that enthalpy extremely decreased (8.1 - 4.0mJ/mg) until 5h, but Tp practically did not change (63.7℃ - 62.6℃) during the time. After 20h, Tp dropped to 60.8℃.

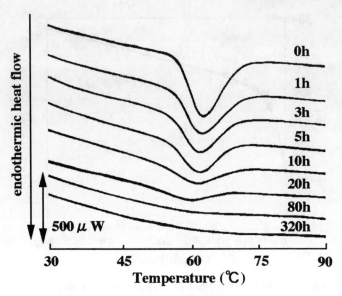

Figure 5 *DSC Patterns of the Ballmill Treated Potato Starch*

Table 2 *DSC Properties of the Ballmill Treated Potato Starch*

Ballmill Time(hr)	To(℃)	Tp(℃)	Tc(℃)	ΔH(mJ/mg)
0	56.3	63.7	73.6	8.1
1	56.3	63.5	73.6	6.2
3	55.8	63.0	72.9	5.7
5	55.4	62.6	72.5	4.0
10	52.3	62.4	72.4	3.1
20	51.0	60.8	71.6	2.4
80	50.3	59.2	69.2	1.4
320	49.5	58.8	66.5	0.9

GPC results of the samples are shown in Table 3. The result of iodine-reaction of sample is shown in Table 4. Separation of amylose fraction from amylopectin was not good. However, λ max decreased as following to amylose fraction area increased by ball-mill treatment. But, the speed of increment in ratio rather small until 20h. Obviously, amylopectin was largely decomposed at 80h, and finally, the fraction was almost disappeared and only amylose fraction remained at 320h. This observation is looked as if only amylopectin is decomposed, but it is proved that amylose is also decomposed from the result of shift of amylose peak in GPC patterns of 320h . From Table 4, it is estimated that 52% of amylose was decomposed at 320h.

Table 3 *Ratio of Ballmill Treated Potato Starch Components Separated by GPC*

Toyopearl HW-75			Toyopearl HW-65		
Ballmill Time	F1	F2	Ballmill Time	F1	F2
0h	70 (100)	30 (100)	0h	89 (100)	11 (100)
10h	67 (96)	33 (110)	10h	85 (96)	15 (136)
20h	62 (89)	38 (127)	20h	81 (91)	19 (173)
40h	52 (74)	48 (160)	40h	72 (81)	28 (255)
80h	35 (50)	65 (217)	80h	52 (58)	48 (436)
320h	6 (9)	94 (313)	320h	13 (15)	87 (791)

Table 4 *Iodine Color Reaction of Ballmill Treated Potato Maize Starch*

Starch Solution

Ballmill Time (hr)	Iodine color reaction			
	λ max (nm)	ε 680	ε 650	ε 600
0	603	0.236 (100)	0.253	0.270
20	598	0.165 (70)	0.178	0.194
80	592	0.147 (62)	0.160	0.180
320	588	0.114 (48)	0.130	0.149

GPC of debranched sample is shown in Table 5. Although change of physiological properties are so large, change of peak area of 3 fractions is very small until 20h. This indicates that break down of chain in amylopectin is not so much. But, consideration that B-2 chain would be selectively disrupted is induced from the results of Table 3. However over 80h, amylose and B-1 chain are also decomposed. At 320h, the shift of peak of No.3 fraction suggests that A chain would be decomposed.

These observations suggest that ball-mill treatment would bring first disruption of hydrogen bond between chains of amylopectin and after, break down of the longer chain and finally, amylose and the shorter chain.

Table 5 *Ratio of Ballmill Treated Potato Starch Components Separated by GPC after debranching*

Ballmill Time(h)	F1	F2	F3
0	21	26	53
5	21	28	51
10	20	30	50
20	19	27	54
80	15	28	57
320	7	23	70

To prove this process, HPAEC was done concerning ball-milled sample of 320h after debranching and removing amylose by Schoch's method. The result of HPAEC which is corrected with sensitivity of glucosidic chain length is shown in Fig.6. Existence of below malthexaose in native is negligible, but existence of malttriose and the above maltoligomer is obvious in sample of 320h. Fig.6-b indicated that disruption of all chain including A chain.

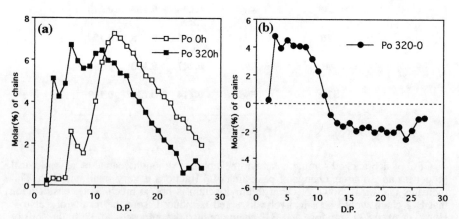

Figure 6 *HPAEC distributions of debranched Potato Starch*
  (a) Potato 0h, Potato 320h
  (b) Potato 320h-0h

It is concluded that ball-mill treatment first attack hydrogen bond of starch, especially amylopectin crystal region and as the result, the starch is gelatinised. And more treatment induces preferential disruption of longer chain (B-2 chain) in amylopectin and next long chain (B-1 chain) and amylose. But , long treatment brings disruption of A chain. These conclusions suggest that amylose would exist in isolated situation from amylopectin crystal region in potato starch. Because, if amylose make crystal region with amylopectin, damage of amylose by ball-mill should be more large.

We also tried to elucidate the component of the outer part and inner part of starch granule by application of ball-mill treatment conbinded with amylase treatment. That is, limited treatment would cause limited gelatinization from outer part, and intact part would appear by amylase treatment. However, this attempt was failed with heterogeneity of starch granule figures and size. Although, four times step-wise treatment were done, and until 3 step, X-ray diffraction patterns and amylose fraction of each step sample are resembled. This observation suggests homogeneity of components of granule.

References

1) V. F. Meuser, R. W. Klingler, and E. A. Niediek starch/stärke 1978, 30, 376-384.

2) W. R. Morrison, R. F. Tester and Gidley Journal of Cereal Science 1994, 19, 209-217.

3) W. R. Morrison, M. J. Gidley, M. Kirkland and J. Karkalas Journal of Cereal Science 1994, 20, 59-67.

4) W. R. Morrison and R. F. Tester Journal of Cereal Science 1994, 20, 69-77.

5) J. Adler, P. Baldwin, and C. D. Melia starch/stärke 1994, 46, 252-256.

THE ROLE OF MOLECULAR WEIGHT IN THE CONVERSION OF STARCH

John R. Mitchell, Sandra E. Hill, Lorna Paterson, Baltasar Vallès, Fiona Barclay
and John M. V. Blanshard

The University of Nottingham
Department of Applied Biochemistry and Food Science
Sutton Bonington Campus
Loughborough, Leicestershire
LE12 5RD

1 INTRODUCTION

The concept of starch conversion is central to its function in a wide range of processes. We
can regard starch conversion as a continuum ranging from the intact crystalline starch
granule to monomeric glucose. The behaviour of starch in an excess water environment at
different levels of conversion is shown schematically in Figure 1. The degree of conversion
that is achieved by a process is determined by heat and mechanical energy input and the
chemical and enzymatic environment.

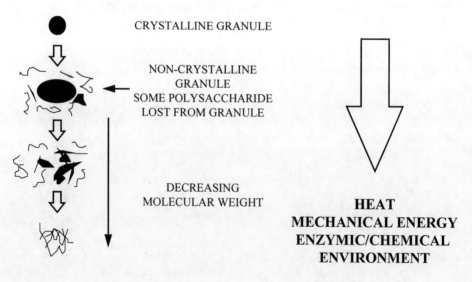

CRYSTALLINE GRANULE

NON-CRYSTALLINE
GRANULE
SOME POLYSACCHARIDE
LOST FROM GRANULE

DECREASING
MOLECULAR WEIGHT

**HEAT
MECHANICAL ENERGY
ENZYMIC/CHEMICAL
ENVIRONMENT**

Figure 1 *The behaviour of starch with increasing levels of conversion in excess water.*

Different food products have different degrees of starch conversion. These cannot be distinguished simply by monitoring order by, for example, birefringence or X-ray methods. Products that contain fully gelatinised starch as judged by these techniques can have very different behaviour. A good example is breakfast cereals that have been processed conventionally by pressure cooking, compared with extruded cereals. Even though the latter have been subjected to processing at high temperatures for much shorter times than their conventional counterparts the degree of conversion is greater because of the greater mechanical energy input.[1,2] This difference in conversion can be measured by determining the water solubility index which is a measure of the amount of polysaccharide found outside the granule, the water adsorption index which is the volume occupied by the swollen granular phase and also by the molecular weight difference as monitored by measurements of the intrinsic viscosity of the solubilised starch.[3]

The role of molecular weight in determining behaviour post gelatinisation is of some interest. It is well established that one of the factors governing granule swelling is the level of bound lipid present. There is also evidence that the pasting behaviour relates to the molecular size of the starch polysaccharides.

One of the questions we consider to be of some interest is whether simple additives with a redox potential used in the processing of starch have an effect on the molecular weight of the polysaccharide component. Initially, we carried out a series of experiments to investigate the role of the polar antioxidant sodium sulfite on granule and polysaccharide integrity.[4,5]

2 RESULTS

2.1 Cassava Starch

Cassava starch was obtained from the Central Tuber Crop Research Institute Trivandrum, India. Sodium sulfite, *n*-propyl gallate, EDTA, and sodium sulfate were purchased from Sigma Chemical Company and were of analytical grade.

    2.1.1 Swelling and Solubility of 1 % Pastes When 1 % cassava starch (variety M4) was pasted under mild stirring conditions and then held at 95 °C for 30 minutes in the presence of varying levels of sodium sulfite a minimum in the swelling volume and a maximum in the amount of polysaccharide detected in the supernatant phase was observed at a sulfite inclusion level of 0.01 %.[4] The addition of propyl gallate, the most polar of the food allowed phenolic antioxidants nullified the effect of the sulfite.

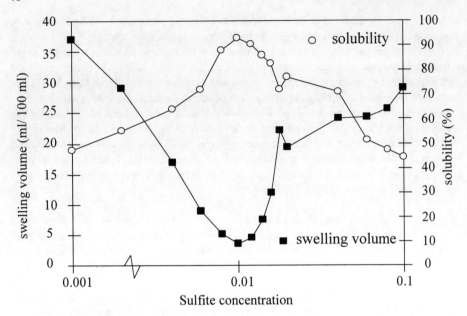

Figure 2 *The swelling volume (■) and solubility (○) of pasted 1 % cassava starch solutions with varying sulfite concentrations.*[4]

2.1.2 *Intrinsic Viscosity* Table 1 shows the intrinsic viscosities of these pasted starches. To determine this the starches were solubilised by adding 1 ml of 5 M KOH to 9 ml of pasted material. It can be seen that the intrinsic viscosity decreased substantially when pasting was carried out in the presence of 0.01 % sulfite. The reduction was less when the sulfite level was 0.1 % and was insignificant when propyl gallate was included in addition to the 0.01 % sulfite.

water	0.01 % Na_2SO_3	0.1 % Na_2SO_3	0.01 % Na_2SO_3 + 0.05 % propyl gallate
182	118	158	176

Table 1 *The intrinsic viscosities ([η] (ml/g)) of cassava starch pasted at 95 °C in the appropriate solvent and then solubilised in 0.5 M KOH*

2.2 Possible Mechanism

Since the intrinsic viscosity of the solubilised material is a measure of its molecular size the data in Table 1 and Figure 2 can be interpreted in terms of a sulfite induced degradation of the starch polysaccharides. The observation that it is prevented by the presence of low levels of an antioxidant would suggest that the degradation is oxidative in nature. At low oxygen levels the interaction between dissolved oxygen and metal ions gives rise to oxygen free radicals that then react with sulfite ions giving rise to superoxide and hydroxyl radicals which subsequently attack the polysaccharide.[4] A possible mechanism is shown in Scheme 1.

The recovery at higher sulfite levels can then be understood in terms of the oxygen scavenging role of sulfite. Some support for this interpretation comes from the observation that if oxygen is bubbled through the solution the recovery seen at high sulfite levels is not observed.[5]

Initiation:

$$M^{n+} + O_2 \rightarrow M^{n+1+} + {}^{\bullet}O_2^{-}$$

Propagation:

$$SO_3^{2-} + {}^{\bullet}O_2^{-} + 3H^{+} \rightarrow HSO_3^{\bullet} + 2\,{}^{\bullet}OH$$
$$SO_3^{2-} + {}^{\bullet}OH + 2H^{+} \rightarrow HSO_3^{\bullet} + H_2O$$
$$HSO_3^{\bullet} + O_2 \rightarrow SO_3 + {}^{\bullet}O_2^{-} + H^{+}$$

Termination:

$$HSO_3^{\bullet} + {}^{\bullet}OH \rightarrow SO_3 + H_2O$$
$$2\,HSO_3^{\bullet} \rightarrow SO_3 + SO_3^{2-} + 2\,H$$
$$SO_3 + H_2O \rightarrow SO_4^{2-} + 2\,H^{+}$$

Scheme 1 *A possible mechanism for the formation, propagation and termination of radicals with the potential to depolymerise starch.*

2.3 Viscosity of 5 % Cassava Starch

Changes in the volume occupied by the swollen starch granule would be expected to have a significant effect on the viscosity of the starch pastes. To determine whether this was indeed the case 5 % cassava starch was pasted at 90 °C under low shear conditions in pH 7.0 phosphate buffer and held at 90 °C for 30 minutes. The sample was then subjected to a high degree of shear with a Silverson rotor-stator mixer. The original objective of this was to try and determine if intact swollen granules were necessary in order for the sulfite effect to be observed. Following pasting, the additives were added in solid form and the paste immediately cooled to 60 °C. A temperature of 60 °C is frequently chosen for rheological measurements on starch systems since complications with retrogradation are avoided. Viscosity was determined as a function of time at 25 °C. Flow curves were obtained over the shear rate range 1 to 70 s^{-1} using a Bohlin CS10 constant stress rheometer equipped with cone and plate geometry (cone angle 4°). Figure 3 displays some representative data from this series of experiments. The results are quoted at a shear rate of 27 s^{-1}.

At these higher starch concentrations, sulfite had the greatest influence at a concentration of 0.1 %, though highly significant differences between the sulfite system and the control were seen at the low 0.01 % sulfite concentrations. Other salts had no influence and complete protection was observed in the presence of propyl gallate. In other experiments it was shown that the chelator EDTA had some protective effect though this was much less than that observed with propyl gallate addition.

Interestingly it was found that sulfite appeared also to act if the pasted starch was held at 30 °C. In the absence of sulfite the viscosity increased with time whereas in its presence it decreased. The observation that the postulated degradation can occur at low temperatures following pasting suggested that an enzymatic mechanism could be a possibility. No difference was observed, however, in the presence of low levels of silver nitrate; therefore although this possibility cannot be ruled out we think it less likely than the oxidative mechanism postulated above.

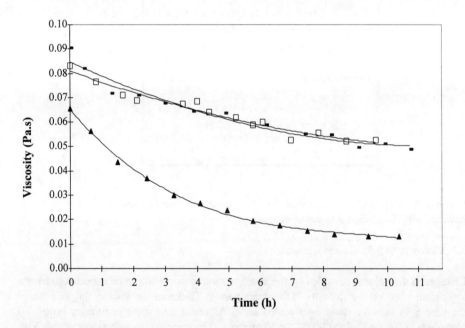

Figure 3 *The viscosity of 5 % cassava starch as a function of time. Starches were pasted in phosphate buffer pH 7.0 at 90 °C. Solids were then added and the change in viscosity followed with time at 60 °C. ■ - control, ▲- 0.1 % sulfite, ☐ - 0.1 % sulfite + 0.01 % propyl gallate. Viscosities reported at a shear rate of 27 s^{-1} and measured at 25 °C.*

2.4 Other Starches

An obvious question is whether other starches behave in a similar way to cassava. Figure 4 shows the influence of sulfite on the swelling and solubility of a range of different starches.

It can be seen that when pasted at 95 °C only sago starch behaves in a directly similar way to cassava starch with a substantial reduction in swollen volume in the presence of 0.01 % sulfite and a recovery when propyl gallate is added. In all cases however there is a significant increase in the amount of material detected in the supernatant (soluble fraction) on sulfite addition that is reduced on propyl gallate addition.

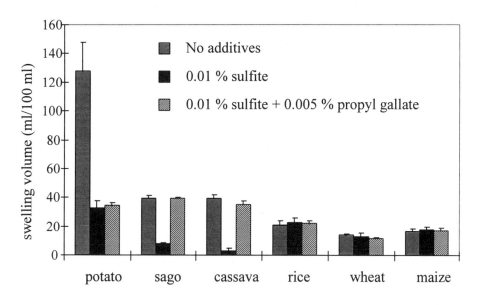

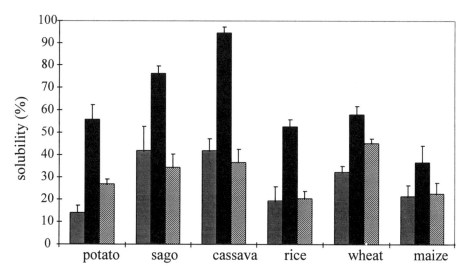

Figure 4 *The influence of sulfite on the swelling volume and solubility of other starches. Data obtained in the same way as the results displayed in Figure 2.*[5]

2.4.1 Potato Starch   Potato starch is a special case since its swollen volume following gelatinisation shows a very large reduction when low levels of NaCl are added. Thus the large decrease in swollen volume on sulfite addition can be interpreted simply in terms of a general electrolyte effect and not degradation. This is seen when the viscosity of 0.9% suspensions are monitored as a function of both sulfite and chloride addition. A similar dramatic decrease in viscosity is found in both bases that parallels the decrease found in the swollen volume (note seen when sodium chloride is added to other starches), as shown in Figure 5a. Nevertheless, when potato starch is solubilised in alkali following pasting a clear difference in viscosity can be seen between potato samples pasted in sulfite and those pasted in a chloride solution (Figure 5b).

3 DISCUSSION

The results presented here confirm that low levels of sulfite can degrade starch polysaccharides as evidenced by both reduced granule integrity, decreased intrinsic viscosities and decreased paste viscosities. Detailed experiments with varying sulfite concentration, propyl gallate, EDTA and oxygen saturation suggest that the starch polysaccharides are degraded by an oxidative reductive depolymerisation mechanism. At low levels of sulfite a prooxidant role is seen for the sulfite. As the levels of sulfite are increased, the level of starch degradation becomes smaller as the sulfite scavenges the system's oxygen.

It is suggested that the molecular weight of the starch polysaccharide is important in controlling granular integrity and hence functionality during processing. The evidence presented above suggests that the molecular weight may be strongly influenced by oxidative reductive depolymerisation. Starch may be more susceptible to this means of degradation than other polysaccharide systems. One possible reason for this is the large size of the amylopectin molecule compared to other polysaccharides. Even if the probability of any given linkage being affected is low; a substantial number of events will still be observed in a very large molecule. The "oxidative" environment would be expected to be influenced by low levels of other components in addition to sulfite, *e.g.* metal ions, lipids and other oxidising agents. For example, materials used as bread improvers such as ascorbic acid may be expected to act partly through their influence on the starch. This possibility is being investigated.

The changes in granular integrity due to molecular weight change may provide a partial explanation for the changes in starch products and cereals upon storage, as well as the differences in pasting and viscosity parameters of apparently similar starches.

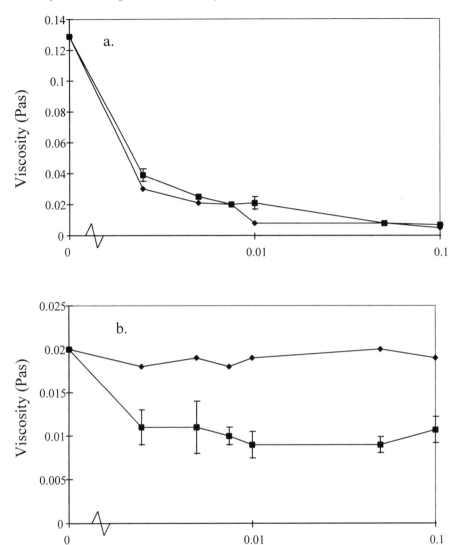

Sulfite/chloride concentration (%)

Figures 5a and 5b *The viscosity of 0.9 % potato starch after pasting in solutions of sulfite* (■) *and chloride* (◆). *Figure a shows the viscosity of a suspension of potato starch, whereas b shows the behaviour of the paste after solubilisation in 0.5 M KOH. Measurements were made at 23 °C and reported at a shear rate of 10 s^{-1}.*

References

1. R. C. E. Guy, "Chemistry and Physics of Baking", ed. J. M. V. Blanshard, P. J. Frazier and T. Gaillard, Royal Society of Chemistry Publication, No. 56, 1986.
2. P. Colonna, J. Tayeb and C. Mercier, "Extrusion Cooking", ed. C. Mercier, P. Linko and J. M. Harper, St. Paul, Minnesota, AACC, 1989, 247.
3. L. L. Diosady, D. Paton, M. Rosen, L. J. Rubin and C. Athanassoulias, *J. Food Sci.*, 1985, **50**, 1697.
4. D. B. Mat Hashim, S. N. Moorthy, J. R. Mitchell, S. E. Hill, K. J. Linfoot and J. M. V. Blanshard, *Stärke*, 1992, **44**, 471.
5. L. A. Paterson, D. B. Mat Hashim, S. E. Hill, J. R. Mitchell and J. M. V. Blanshard, *Stärke*, 1994, **46**, 288.
6. L. A. Paterson, J. R. Mitchell, S. E. Hill, and J. M. V. Blanshard, submitted to *Carb. Res.*

DIFFUSION AND REACTION IN WHEAT GRAINS

A.G.F. Stapley, P.J. Fryer and L.F. Gladden*

School of Chemical Engineering, The University of Birmingham, Edgbaston, Birmingham, B15 2TT
*Department of Chemical Engineering, University of Cambridge, Cambridge, CB2 3RA

1 INTRODUCTION

The steaming and boiling of wheat grains is the basis for much commercial cereal processing. The cooking of the grains is usually one of the first steps in the manufacture of breakfast cereals. After processing, the grains are soft enough to be mechanically worked to form the shape of the final breakfast produce, such as rolling into flakes. Both boiling and steaming involve two major processes: *diffusion of water* into the grain, followed by *gelatinisation (or conversion) of the raw starch* within the wheat grain into an edible form due to the combined presence of heat and moisture. A great deal of research has been directed at the gelatinisation of powdered starch/water systems[1]. It is now generally acknowledged that it involves an interplay between two processes: melting of starch crystallites and a glass transition in the amorphous regions[2].

Only a few studies have been published on whole grain systems (mostly for rice). Diffusion studies have largely been based upon external mass balance data of moisture uptake from which interpretations have been made. However, variations in the size and shape between grains, and the peculiar shape of starch grains makes it difficult for any real conclusions to be made.

In this paper, we have employed magnetic resonance imaging (MRI) to study diffusion. MRI is a non-invasive and non-destructive technique, which can map the distribution of water protons on a particular plane within a sample. MRI is being used in an ever increasing number of applications, most notably in the medial field as a powerful non-invasive diagnostic and research tool. It is also finding applications in food systems[3]. Here, we have used MRI to image the water distribution on a plane within wheat grains at various stages of cooking. A useful feature of the technique is that it does not detect molecules that are in a "solid" state as nuclei which are held stationary relax several orders of magnitude faster than the time scale of the experiment. This is useful here as the signal arising is almost exclusively due to the liquid water in the grain, and is not confused by hydrogen nuclei from the starch matrix. Combined with gravimetric data, one can obtain virtually quantitative maps of water content. There is thus no need to infer models from external mass balance data - you can actually see the distribution. MRI is used here to study wheat grain cooking in two different environments of steam and liquid water at atmospheric pressure (c. 100°C) and also at 2 bara (c. 120°C). Only data from experiments at 1 bara are presented here.

In addition to MRI, we have employed differential scanning calorimetry to scan grains processed in the same way as in the MRI experiments. The DSC peaks resulting are due to starch crystallites that remain unmelted after cooking. DSC has been widely used to characterise the gelatinisation of powdered starch/water mixtures, but is used here with intact grains. The use of both techniques in tandem offers potentially powerful tools in aiding our understanding of grain cooking, which may ultimately lead to improved operating strategies in cereal manufacture.

2 MATERIALS AND METHODS

2.1 Sample Preparation

Samples of Riband variety grains (of 'mealy' type) of mass between 35 and 40 mg were cooked at 1 bara (c. 100°C) or 2 bara (c. 120°C) in a steam pressure cooker fitted with a pressure gauge. Distilled water was brought to boiling at atmospheric pressure before insertion of the samples. Samples for steaming were suspended on a wire mesh above the water, whereas samples for boiling were placed in small pans of water at the bottom of the pressure cooker. Nominal cooking times for 2 bara samples were measured from the time of insertion of the samples to the time at which depressurisation commenced. Rapid depressurisation was achieved by venting for boiled samples (60 - 80 secs), but was allowed to occur by natural cooling for steamed samples (170 - 205 secs). Boiled samples were then plunged into cold water for 30 seconds to achieve rapid cooling.

In addition, for DSC experiments, grains were soaked for 2-3 days, dried to the requisite moisture content, and then stored overnight in small plastic vials to equilibrate. A few soaked grains were found to have begun to germinate, and were discarded.

Moisture contents of the grains were obtained by weighing before and after treatment; increased mass was attributed to absorbed water. It was assumed that all raw grains possessed an initial water content of 12%, from vacuum drying grains of the same batch at 50°C for 9 days. Cook times were; 5 - 90 minutes (steamed samples), 5 - 30 minutes (boiled samples).

2.2 Magnetic Resonance Imaging

Cooked grains were imaged in two sets of experiments using a Bruker MSL200 and a Bruker DMX200, both with micro-imaging systems. The imaging probe inserts had an internal diameter of 5 mm. Except where specified the spin-echo technique was used with an echo time (T_E) of 2.2 ms and a repetition time (T_R) of 2 s. Transverse slices (~0.6 mm thickness) were taken through the centre of the grains (across the crease). Gradients in read- and phase-encoding directions were typically 7 G cm^{-1} (MSL) and 40 G cm^{-1} (DMX). NMR experiments were also performed to determine bulk longitudinal (T_1) and transverse (T_2) relaxation times to assess to effect of T_2 and T_1 weighting on the images. All grains were imaged within 3 hours of cooking.

It was found that T_2 relaxation would have a significant weighting effect on the images - favouring areas of highest moisture content. It was thus necessary to correct the images to provide quantitative maps of moisture content. A linear calibration between observed pixel intensity and moisture content using average pixel intensity data from the more homogeneous images was found to be most reliable[4].

2.3 Differential Scanning Calorimetry

Prepared grains were inserted into Perkin Elmer Large Volume Stainless Steel DSC capsules (75 µl), which were then immediately sealed. Sealing the capsules allows the grains to be scanned to above 100°C, as evaporation is limited to the capsule volume and to the saturated vapour pressure corresponding to the current sample temperature. The capsules were then inserted into a PC-driven Perkin-Elmer DSC-7 differential scanning calorimeter, with an empty sample pan acting as the reference sample. The sample was then scanned at a rate of 10°C min^{-1} from 10°C to 200°C. This scanning rate was chosen as it gives a reasonable peak height, and errors due to thermal lag are not excessive.

3 RESULTS & DISCUSSION

Results are presented for processing at 1 bara (100°C); results for processing at 2 bara are reported elsewhere[4].

3.1 Magnetic Resonance Imaging

The images of grains obtained at various stages of boiling at 1 bara are presented in Figure 1. They show water penetrating evenly into the wheat grain from all points on the boundary including the inside of the crease. There is also a very well defined "front" where the moisture content drops from a high moisture content around the exterior of the grain to a lower moisture content in the core over a comparatively short distance. The front moves to the centre of the grain during cooking. Similar behaviour is also obtained at 2 bara (not shown) over the course of 15 minutes.

The concentration profiles obtained are similar to profiles of iodohexane diffusing in polystyrene measured using Rutherford backscattering spectrometry[5]. These show some characteristics of Case II diffusion whereby the relaxation of polymer chains controls the ingress of the solute. Pure Case II diffusion shows a very sharply defined front which delimits an outer rubbery region from an unpenetrated glassy core. Both the iodohexane/polystyrene and water/wheat grain systems show a less dramatic front which can be classified as 'anomalous diffusion' - intermediate between Fickian and Case II diffusion. This suggests that polymer relaxation at least partially controls diffusion.

Images of grains steamed at 1 bara (Figure 2) show much reduced water contents (confirmed by gravimetric data). There is no indication of a "front" and the distribution of water in the grains is much more homogeneous. The level of moisture in the grains slowly increases with cooking time. There is conceivably a thin layer of high moisture at the surface, which is probably caused by condensation at the beginning of steaming as the grain warms up to the cooking temperature. Grains steamed at 2 bara also show a more homogeneous distribution, although higher moisture contents were consistently found at the centre of the grain. This may be due to flashing off of moisture from the exterior parts of the grain during depressurisation.

The homogeneity of the steamed grain images and the generally slower uptake of moisture suggests a mass transfer resistance for steaming which is not present in boiling. This resistance may be either at the surface of the grain (i.e. the pericarp or a thin layer of gelatinised starch), or at the starch granule level where polymer relaxation may again control the process (steam being allowed to penetrate within the grain through interstices between starch granules).

Heat transfer considerations must also be taken into account. As steam condenses onto (or within) the grain, latent heat is released which raises the grain temperature above that of the surrounding steam. This causes convective heat loss from the grain to the steam, which balances the latent heat production. The temperature rise has been confirmed experimentally by thermocouple experiments, with differences of 2-3°C being observed. The rise in grain temperature has a depressing influence on the thermodynamic driving force for mass transfer which may partially account for the slower uptake of moisture.

3.2 Differential Scanning Calorimetry

A number of peaks were observed for soaked grains. Conceivably 5 endotherms exist although some may be thermal events not connected with starch transitions. The peak temperatures are shown in Figure 3 (marked I - V). They match well with literature data for powdered starch/water mixtures[6] (marked i - iii), suggesting that starch within wheat grains converts at a similar temperature as powdered starch at the same moisture content. This confirms that water content is the most important factor in determining starch transition temperatures in whole grains.

Boiled and steamed grains showed different DSC results. Steamed grains showed a clearly defined peak, the temperature of which falls as cooking proceeds and the moisture

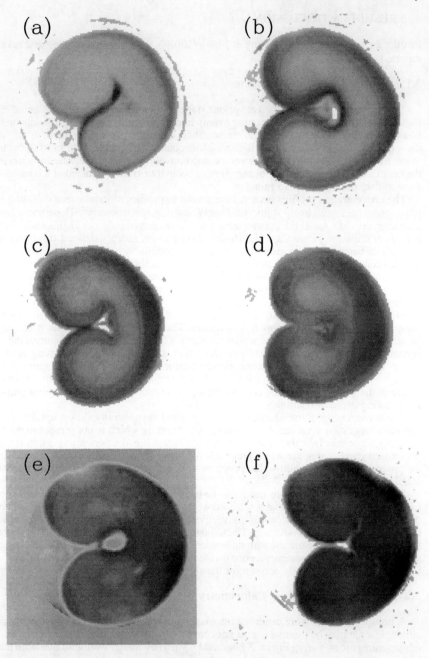

Figure 1 *MRI images of moisture content for grains boiled in distilled water at 1 bara (c. 100°C) for (a) 5 minutes, (b) 10 minutes, (c) 15 minutes, (d) 20 minutes, (e) 30 minutes. The moisture content scale runs from 0% (white) to 70% (black). The pixel resolution is 37 μm.*

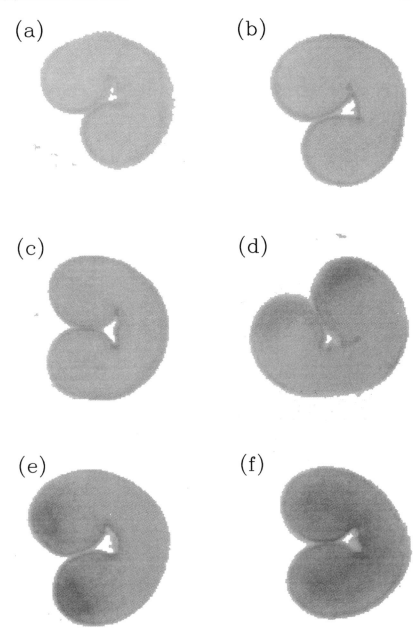

Figure 2 *NMR images of moisture content for grains steamed using distilled water at 1 bara (c.100°C) for (a) 5 minutes, (b) 15 minutes, (c) 30 minutes, (d) 45 minutes, (e) 60 minutes, (f) 90 minutes. The moisture content scale runs from 0% (white) to 70% (black). The pixel resolution is 37 μm.*

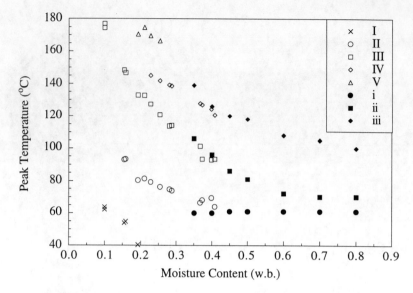

Figure 3 *Peak temperature vs moisture content for raw and soaked grains (I, II, III, IV, V), and comparison with literature data for powdered starch (i, ii, iii - data from Eliasson, 1980).*

content rises. The DSC peak temperatures of steamed grains mirrored those of uncooked grains at equivalent moisture contents, except at higher moisture contents where significant "clipping" of the endotherms occur. The onset temperature, however, was consistently at a point approximately 7°C above that of the steaming temperature. Starch granules with a transition temperature below the cooking temperature, it appears, have already undergone conversion within the cooker. That the onset does not exactly correspond to the steaming temperature can be partly attributed to thermal lag in the DSC, and partly due to the grain temperature rising above that of the steam temperature during cooking.

Boiled grains produced broader endotherms than the steamed grains, the peak temperatures of which varied little with moisture content. The difference can be explained with reference to the MRI images, which show a larger range of moisture contents for boiled grains than for steamed grains. The disappearance of the endotherm is on the same time scale as the movement inwards of the moisture front, and it is therefore believed that the front corresponds to a gelatinisation zone.The DSC signal for boiled grains arises from an unreacted central core (ahead of the moisture front), with the starch lying behind the front totally converted and making no contribution. This ties in well with the MRI imaging results which suggest that polymer relaxation (i.e. a glass transition) influences the diffusion rate. Starch gelatinisation is widely acknowledged to involve a glass transition step[1,2].

4 CONCLUSIONS

The combined MRI and DSC evidence point to two different mechanisms for boiling and steaming. For boiling, MRI suggests that the diffusion of water into the grain is influenced by polymer relaxation (or glass transition), which gives rise to an incoming diffusion front. Such a view is consistent with DSC evidence which indicates a gelatinisation front (which involves a glass transition) that moves into the grain at a similar speed. MRI images of

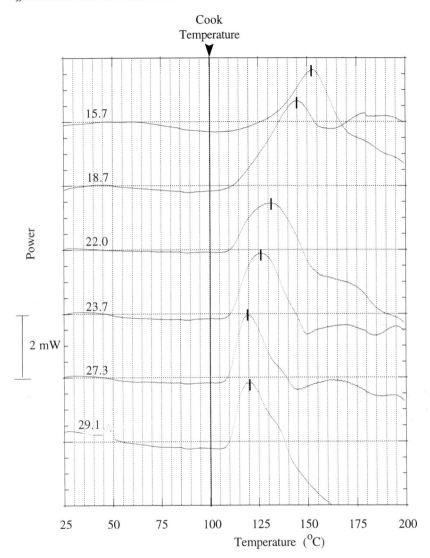

Figure 4 *DSC endotherms of grains steamed at 1 bara (circa 100°C) for (from top to bottom) 5, 15, 30, 45, 60 & 90 minutes. Values shown are mean percentage moisture contents (wet basis).*

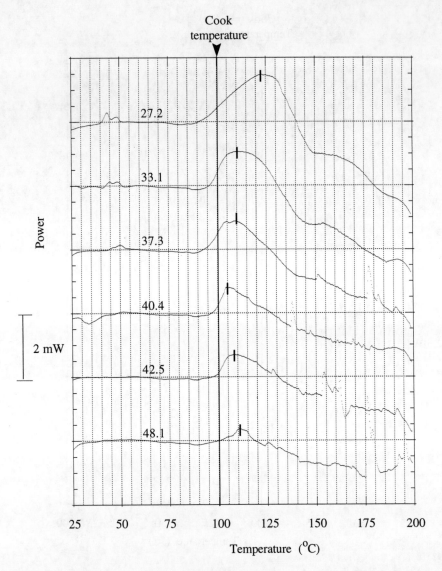

Figure 5 *DSC of grains boiled at 1 bara (circa 100°C) for (from top to bottom) 5, 10, 15, 20, 25, & 30 minutes. Values shown are mean percentage moisture contents (wet basis).*

steamed grains showed much more homogeneous images, and a slower uptake of water, suggesting the presence of a mass transfer resistance not observed for boiling. The nature of this resistance cannot be elucidated from these experiments. The steaming process may be heavily influenced by heat transfer considerations as the latent heat of condensing steam needs to be convected away from the grain, resulting in the grain temperature rising above that of the surrounding steam. Starch conversion in steamed grains occurs when the moisture content has risen high enough for the DSC endotherm to dip below the grain temperature.

Acknowledgements

We wish to acknowledge Weetabix Ltd. and the BBSRC for funding the project and to Weetabix Ltd. for useful discussions.

LFG wishes to thank the Process Engineering Committee of the EPSRC for the award of the MRI spectrometers.

Many thanks to Mike Hollewand and Thomas Hyde for the acquisition of the MRI images.

References

1. C.G. Biliaderis, In: Thermal Analysis of Foods, Eds. V.R. Harwalkar and C.Y. Ma, 1990, Elsevier Applied Science, London, Chapter 7, p.168.
2. L. Slade and H. Levine, *Carbohydr. Polym.*, 1988, **8**, 183.
3. G.W. Schrader, J.B. Litchfield and S.J. Schmidt. *Food Tech.*, 1992, **46**, 77.
4. A.G.F. Stapley, PhD Thesis, University of Cambridge, 1995.
5. C.Y. Hui, K.C. Wu, R.C. Lasky and E.J. Kramer, *J. Appl. Phys.*, 1987, **61**, 5137.
6. A.C. Eliasson, *Starch/Stärke*, 1980, **32**, 270.

THE EFFECTS OF WATER AND SUGARS UPON THE RETROGRADATION OF AMYLOPECTIN IN THE REGION (T-T_g) BETWEEN 0 AND 85°C

The University of Nottingham
Department of Applied Biochemistry and Food Science
Sutton Bonington Campus
Loughborough, Leicestershire
LE12 5RD

Imad A. Farhat, John M.V. Blanshard, Jennifer L. Melvin and J R Mitchell

1. INTRODUCTION

The restoration of crystalline order in gelatinized starch systems, frequently referred to as retrogradation, or recrystallization, is a problem which has received intensive attention over the past 70 years because of both its commercial importance and also the scientific questions relating to its nature and control [1,2,3,4,5,6]. It has also been the subject of several substantial reviews [7,8]. However, the majority of the scientific research which has sought to gain an understanding of the relevant molecular events occurring during retrogradation has focused on systems with water contents of >80% H_2O (dry solids basis, d.s.b). More recently the importance of the glass rubber transition in understanding the process and the application of concepts well known in polymer physics [8,9] have led to a number of systematic studies on the behaviour of such systems at lower water contents (<80% d.s.b) [1,2,3,6,7], including sub-T_g transitions. This paper will review recent studies [11] from our laboratory which address the roles of water and sugars in the retrogradation of gelatinized waxy maize starch.

A variety of techniques have been employed to follow starch retrogradation, but particular attention will be placed in this report on the use of low resolution, pulsed NMR and wide angle x-ray diffraction. The literature already contains the results of two studies [12,13] which have used pulsed NMR to follow the change in the free induction decay (FID) with time. In so doing, they monitored the combined behaviour of protons from the 'solid' polymer matrix and the complex 'liquid' components. In the current work, the solid and liquid components in the FID were deconvoluted but, in addition, the mobility of the water protons was followed by the quite different Carr-Parcell-Meiboom-Gill (CPMG) procedure. All of these three measurements might be expected to show distinct changes with progressive retrogradation. Changes in the mobility or amplitude of the water ('liquid') component may also have implications for changes in perceived texture.

In contrast to the use of NMR, wide angle X-ray scattering (WAXS) provides information on the development of molecular order in domains of > 1nm.

2. MATERIALS AND METHODS:

2.1 Sample Preparation:

Amylopectin as waxy maize starch was supplied by National Starches and Chemicals Co., Manchester, U.K. Sugars were obtained from Sigma Chemical Co. Ltd., UK. The amylopectin-sugar samples were mixed using a Kenwood Peerless planetary mixer for 1 hour prior to extrusion. All materials were extruded through a 1mm x 30mm slit die using a Clextral BC-21 co-rotating, intermeshing twin screw extruder and yielded a non-expanded ribbon. The extrusion temperature profile was 40, 110, 120 and 80°C. Distilled water was pumped into the second zone of the barrel at a level (after taking into account the water content of the raw materials) to give the desired water content. The sample composition was calculated on a total 'dry solids basis' (d.s.b). For example, a 70:30:25 amylopectin/sugar/water system refers to a mix containing 70g amylopectin, 30g sugar (30%) and 25g water (25%). Extrudates containing fructose, sucrose and xylose at two concentrations (10% and 30% w/w d.s.b.) were prepared. Samples were sealed and stored at 40±0.1°C in the spectrometer probehead for the NMR measurements, and at 25±2 °C in a temperature controlled room for the x-ray studies.

2.2 Measurements:

The proton NMR relaxation parameters were measured using a bench top Minispec PC120 (Bruker Spectrospin) operating at 40°C and a resonance frequency of 20 MHz as described in more detail elsewhere [11]. Spin-spin relaxation parameters were obtained from the NMR free induction decay signal (FID) recorded directly after a 90° rf pulse or using the CPMG experiment with a τ spacing of 262 μs between 90° and 180° pulses. The free induction decay was deconvoluted into 2 components, a rigid 'solid' component with decay times ($T_{2\,solid}$) of a few tens of μs and a more mobile 'liquid' component with decay times ($T_{2\,liquid}$) of a few hundreds of μs [11]. The mobility of the observed ^{1}H containing molecular species was the criterion governing their assignment to the rigid or the mobile component of the NMR signal. The CPMG decay was best described by a single exponential.

X-ray spectra were recorded for values of 2θ between 4° and 38° using the APD-15 Philips diffractometer equipped with a copper tube operating at 40 kV and 50 mA producing CuK$_\alpha$ radiation of 1.54 Å wavelength. Crystallization and increased order were marked by the development of narrow spectral features upon the broad background, characteristic of the amorphous state. An increase in the relative amplitude of one of the two peaks was used to quantify the increase in crystalline order. Crystallinity indices (ci) were obtained from the shift corrected intensity of the peak at 17.2° normalized to the intensity of 16° 2θ.

$$ci = (I_{17.2°} - I_{16°}) / I_{16°} \qquad (1)$$

3. RESULTS AND DISCUSSION:

3.1 Analysis of Results:

3.1.1 Modelling Recrystallization:

The process of recrystallization of amylopectin in which crystallites can grow from different nucleation centres can be modelled by the Avrami equation [14] in which, during crystallization, the unconverted polymer fraction at time t, U(t), can be described by :

$$U(t) = \exp(-kt^n) \qquad (2)$$

where k is the rate constant and n, conventionally, is believed to be a function of the type of nucleation [19]. U(t) was calculated from the experimental data using

$$U(t) = \frac{Y_{max} - Y(t)}{Y_{max} - Y_o} \qquad (3)$$

Y(t) is any physical property affected or any parameter related to the crystallization of the material and in this paper refers to changes in the crystallinity index or NMR relaxation parameters. Y_o and Y_{max} are, respectively, the initial (t=0) and maximum plateau values of Y(t). The results were modelled using a least squares minimization fitting routine. The actual equation used was obtained by replacing U(t) by its relationship Y(t) in the Avrami equation.

$$Y(t) = Y_{max} - (Y_{max} - Y_o) \exp[-kt^n] \qquad (4)$$

with 4 adjustable parameters Y_o, Y_{max}, k and n. In general, a satisfactory agreement was found between this model and the experimental results ($\chi^2 < 10^{-4}$).

The rate of retrogradation can be obtained by writing the Avrami equation in a slightly different form

$$U(t) = \exp[-(RT)^n] \qquad (5)$$

where R is the rate of retrogradation and has an inverse time dimension (time^{-1}).

3.1.2 Calculation of T_g:

It is generally accepted in synthetic polymer systems that the process of crystallization is confined to a temperature domain bounded by T_g, the glass transition and T_m, the polymer melting temperature. It should therefore be possible to determine the impact of proximity of T_g to the storage temperatures by calculating the T_g of the experimental systems using the Ten Brinke-Karasz [16,17] equation

$$T_g = \frac{\sum_i w_i \Delta C_{p_i} T_{g_i}}{\sum_i w_i \Delta C_{p_i}} \qquad (6)$$

T_{gi} refers to the glass transition temperatures of the components: water, 134K; amylopectin, 502K; sucrose, 343K; xylose, 286K; fructose, 280K. ΔC_{pi} refers to the changes in specific heat of the components at the glass transition: water, 1.94 Jg^{-1} K^{-1}; amylopectin, 0.41 Jg^{-1} K^{-1}; sucrose 0.76 Jg^{-1} K^{-1}; xylose 0.95 Jg^{-1} K^{-1}; fructose, 0.76 Jg^{-1} K^{-1} [9,18]. T_g is the glass transition of the mixed system and w_i the weight fraction of component i in the mixture. The equation can be used to determine the T_g of both binary (amylopectin-water) and ternary (amylopectin-sugar-water) systems.

3.2 Binary Systems:

3.2.1 WAXS:

Immediately after extrusion there was no evidence of crystalline order. On storage at 25°C, there was a clear recovery of crystallinity. The line shape of the x-ray spectra recorded on the fully retrograded samples depended strongly on the amount of water incorporated in the system during the extrusion process. As the water content increased, these x-ray spectra exhibited patterns increasingly similar to the B-type starch polymorph. This was accompanied by an increase in the intensity of the characteristic peak at approximately 5.8° and a decrease in the line width of this peak at 17.2° whilst the line width of the peak at approximately 23° increased, i.e. a progressive transition from the A to B polymorphic form (fig 1). The results are in agreement with the A/B temperature-water-content polymorph diagram (fig 2) described by Marsh [2].

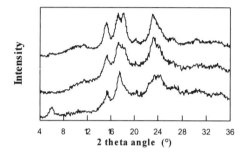

Fig. 1 XRD spectra recorded on, from top to bottom: native waxy maize starch containing 34 % water (A-type), and extruded with 35% and 65% water.

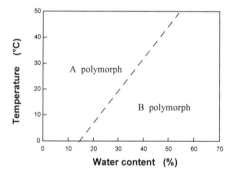

Fig. 2 Temperature-water content diagram for A/B starch polymorphs as described by Marsh (1986) [2]

3.2.2 Proton Relaxation NMR:
The free induction decay (FID) and the spin-echo decay (CPMG) both showed a strong dependence on the duration of storage. The relaxation times of the rigid and mobile components and their relative contributions, were obtained by deconvolution of the FID.
  The results showed a decrease in T_2 of the solid component on storage (fig 3) accompanying a not unexpected increase in the fraction of the proton population contributing the solid component of the FID. These protons, initially part of the FID liquid-like component with a decay time of 100's of μs, became much less mobile with decay times of 10's of μs, resulting in a greater amplitude of the rigid component. There

is therefore a clear decrease in the mobility of the polymer as it progresses through the reordering transition, leading to the conclusion that the gelatinized starch component is significantly more mobile than that in the ordered crystalline fraction, a view supported by other CP-MAS and FT-MAS experiments conducted in this laboratory. The spin-spin relaxation times were obtained by fitting the CPMG decay to a single exponential. The T_2 values representing mainly the water component in the system decreased considerably as the retrogradation of the starch progressed. The direct interpretation of this observation suggests a lower mobility of the water in the retrograded system compared to the freshly gelatinized gel. More realistically the water relaxation is reflecting a dependence upon polymer mobility.

The time dependencies of these changes in the NMR relaxation parameters describing the different molecular dynamics of the various components were comparable for any one sample, but the values obtained depended strongly on the water content present in the sample.

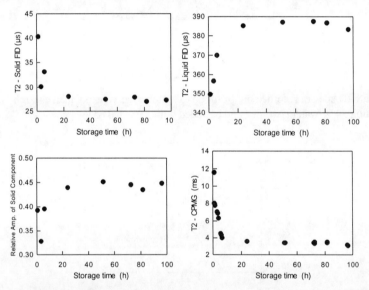

Fig. 3 ^{1}H relaxation NMR results for a waxy maize starch extrudate containing 41% water (w/w d.s.b) for different storage times (40 ± 0.1°C).

3.3 Ternary Systems:

3.3.1 WAXS:

Again the starch granules were found to be fully gelatinized after extrusion and no evidence of sugar crystallization was observed save in the samples containing 30% xylose and less than 24% water which were not considered further for this paper. As with the binary system, crystallinity increased with storage time.

A notable feature was the presence of the so-called pseudo-B-polymorph in fully retrograded samples first described by Marsh [2,3]. The characteristic features of the pseudo-B type X-ray pattern are: the peak at 5.6-5.8° 2θ is much reduced in intensity, if not wholly absent, there is a sharp peak at 17.2° (which is doubled by a second peak at

~18.2 in the A-type) and two peaks are present at approximately 22.3 and 24.2° (which merge into one peak at ~23.2 2θ in the A-type (fig 4).

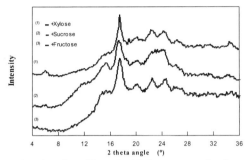

Fig. 4 XRD spectra showing the effect of added sugars on the formation of the pseudo-B polymorph in retrograded waxy maize starch extrudates (initially A type) at 29±1% water.

A number of hypotheses may be postulated to explain the origin of this structure.
(i) the unequal partitioning of water in the system in favour of the starch leading to a larger amount of water effectively associated with the amylopectin - this would not account for the diminution/disappearance of the 5.6° 2θ peak. Furthermore, preferential hydration is more likely to be in favour of the sugars if the sorption isotherms of the various components fairly represent the mixed systems.
(ii) the accommodation of sugar molecules within the cell unit cavity of the B-starch crystalline structure proposed by Imberty *et al.* [19,20]. The accessible domain is of the order of 0.7 nm which could accommodate a sugar molecule.

There is certainly evidence of molecular interactions between the amylopectin component and the sugars from ^{13}C solid state NMR studies which have shown that a certain fraction of the sugar present in the system had a restricted mobility which was detected with the CP-MAS experiment, while the remainder of the sucrose was more mobile and had a solution-type behaviour which was recorded using the FT-MAS technique (fig 5).

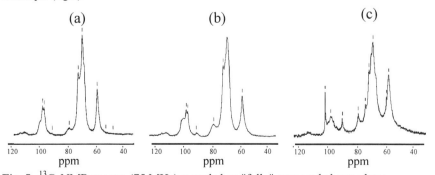

Fig. 5 ^{13}C NMR spectra (75 MHz) recorded on "fully" retrograded extrudates :
(a) CP-MAS amylopectin-water (100:60), (b) CP-MAS amylopectin-sucrose-water (70:30:44) and (c) FT-MAS amylopectin-sucrose-water (70:30:44)

The sharp lines in (c) show the solution type behaviour of a large proportion of the sucrose present in the system, the sharp lines particularly at 80.6, 91.4 and 102.9 ppm in (b) show the contribution of a fraction of sucrose to the signal of the rigid component. (a) is presented for comparison.

3.3.2 Proton Relaxation NMR:
As for the binary system, the FID and CPMG decays recorded on the starch-sugar-water extrudates showed considerable changes similar to the starch-water system. The presence of the sugars affected the retrogradation kinetics of amylopectin as monitored by the NMR relaxation times - storage duration curves (fig 6).

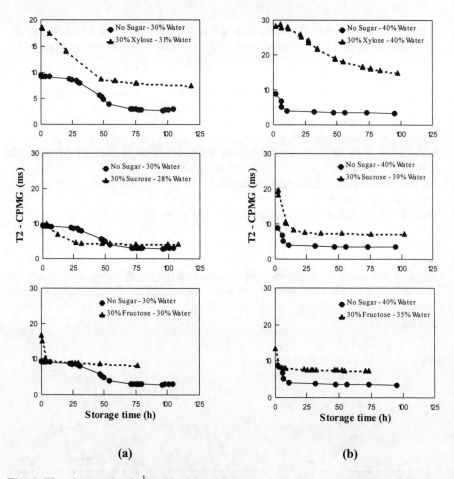

Fig. 6 The changes in the ^{1}H spin-echo relaxation time ($T_{2\ CPMG}$) during storage for waxy maize starch - sugar extrudates (70:30) containing (a) 29±1% water and (b) 40±1% water (except for the fructose containing sample which had 35% water).

4. GENERAL DISCUSSION:

A number of interesting points emerge from these results.

1. Firstly it is clear that the particular polymorph which results from the recrystallization of the starch is a function of the water content and temperature. The temperature of maximum rate of recrystallization is also a function of water content (unpublished results).

2. From an experimental point of view, it is not surprising that a number of NMR parameters give evidence of the development of increased retrogradation. These include the T_2 of the solid and the T_2 of the liquid components of the FID, the relative amplitude of the solid component of the NMR signal, and the T_2 derived from the CPMG measurements. The fact that a sample can be left in the spectrometer probehead during the retrogradation enabling an uninterrupted recording of the process is particularly useful.

3. A detailed study of a whole range of results of which figures 6 (a) and (b) are just examples, shows quite clearly that although the water content, particularly in ternary systems, is a poor indicator of the presence of conditions conducive to retrogradation, replotting the results with $(T- T_g)$ as the abscissa shows that the rate does indeed approach zero at $(T-T_g=0)$ (fig 7). The T_g of each system was calculated using the Ten Brinke and Karasz equations [16,17].

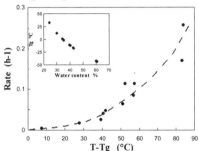

Fig. 7 The retrogradation rate obtained from the Avrami modelling of $1/T_2$ CPMG (40°C) results as a function of $(T-T_g)$. The insert shows the effect of water content on T_g.

4. If we consider the effects of different sugars, a whole range of results have been derived that show that the rate of retrogradation depends on the water content and the particular sugar. Figure 8 summarises the effect of 30% sugars on the retrogradation rate values derived by NMR and x-ray diffraction. The results are notably different and indicate that whereas xylose reduces the rate of retrogradation slightly, sucrose enhances the retrogradation when $(T-T_g)<60$ °C. At higher $(T-T_g)$ values, i.e > 60°C, sucrose would appear to depress the rate. Fructose, under the conditions of the experiment, uniformly enhances retrogradation but, like sucrose, might well reduce retrogradation rates at $(T-T_g)>135$°C. Such conditions could be obtained, either by increasing T, the temperature of storage, or reducing T_g by increasing the water content.

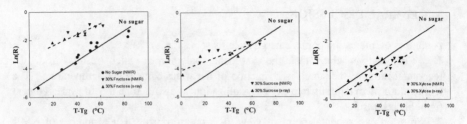

Fig 8 The linear relationship between Ln(R) and the calculated T-T$_g$ for samples containing 30% fructose, sucrose and xylose.

5. It is also a matter of interest to see if these results can be fitted to the WLF [21] equation. Figure 9 shows that this is feasible but yields non-standard values of C$_1$ and C$_2$. Such an equation can only hold over a specified range if that lies within either the negative or positive gradient arms of the inverted rate constant versus T-T$_g$ parabola. We presume the results recorded are in this instance on that portion with a positive gradient.

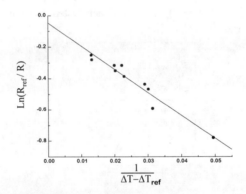

Fig. 9 The WLF [21] type plot of the retrogradation rate as derived from the T$_{2\,CPMG}$ ^{1}H relaxation NMR results of an amylopectin-water extrudates.

5. CONCLUSION:

A number of interesting observations have emerged from these studies which require further investigation. For example, the apparent progressive decreased mobility of water molecules with retrogradation has important implications for the perceived dry texture of staled products.

Conversely, the effect of sugars upon the x-ray spectrum the reduction or even elimination of the 5.6 2θ peak suggests that water is being displaced from the helical structure and possibly made more available as a textural plasticizer.

The exact role of sugars in relation to the recrystallization of the starch is still enigmatic - why some enhance and others depress the rate. We believe this is the result of specific molecular interactions. It is also fascinating that sugars which apparently enhance the rate of crystallization at temperatures just above T_g (sucrose and fructose) display a behaviour which demonstrates (sucrose) or suggests (fructose) that at high values of (T-T_g), they will exert an inhibiting effect upon the reorganization process.

Acknowledgements

The authors wish to thank the ACTIF II (Amorphous Crystalline Transition in Foods) consortium of industrial companies and the MAFF for the financial permitted the completion of this research programme.

REFERENCES:

1. Ring S.G, Colonna P, L'Anson K.J, Kalichevsky M.T, Miles M.J, Morris V.J and Orford P.D (1987), *Carbohydrate Research*, **162**, 277-293

2. Marsh R.D.L, 1986 , *Thesis*, University of Nottingham, 54-78

3. Marsh R.D.L and Blanshard J.M.V (1988), *Carbohydrate Polymers*, **9**, 301-317

4. Wilson R.H, Goodfellow B.J, Belton P.S, Osborne B.G, Oliver G and Russel P.L (1991), *J. Sci. Food & Agric.*, **54**, 471-483

5. Miura M., Nishimura A. and Katsuta K. (1992), *Food Structure*, **11**, 225-236

6. I'Anson K.J, Miles M.J, Morris V.J, Bedford L.S, Jarvis D.A and Marsh R.A (1990), *J. Cereal Sci.*, **11**, 243-248

7. Kulp K and Ponte J.R (1981), in *Critical Rev. in Food Sc. and Nutrition*, **15**, 1-48

8. Slade L and Levine H, (1991), in *Critical Rev. in Food Sc. and Nutrition*, **30**, 115-360

9. Kalichevsky M.T, Blanshard J.M.V (1993), *Carbohydrate Polymers*, **20**, 107-113

10. Wang Y.J and Jane J, (1994), *Cereal Chemistry,* **71**, No 6, 527-531

11. Farhat I.A, Melvin J.L, Blanshard J.M.V, Mitchell J.R and Derbyshire W. (1996), *in press.*

12. Le Botlan D and Desbois P, (1995), *Cereal Chemistry*, **72**, No 2, 191-193

13. Teo C.H and Seow P., (1992), *Starch*, **44**, No 8, 288-292

14. Avrami M, 1941, *J. Chem. Phys.*, **9**, 177

15. Wunderlich B (1976), *Macromolecular Physics*, Academic Press NY, **2**.

16. Couchman P.R and Karasz F.E (1978), *Macromolecules*, **16**, 117-119

17. Ten Brinke G, Karasz F.E and Ellis T.S (1983), *Macromolecules*, **16**, 244.

18. Orford P.D, Parker R and Ring S.G (1990), *Carbohydrate Research*, **196**, 11-18

19. Imberty A, Chanzy H, Pérez S, Buléon A and Tran V (1988), *J. Mol. Biol.*, **20**, 2634 2636

20. Imberty A and Pérez S (1988), *Biopolymers*, **27**, 1205-1221

21. Williams M.L, Landel R.F and Ferry J.D (1955), *J. Amer. Chem. Soc.*, **77**, 3701-3707

MECHANICAL PROPERTIES OF WHEAT FLAKES AND THEIR COMPONENTS

D.M.R. Georget and A.C. Smith,

Institute of Food Research,
Norwich Research Park,
Colney, Norwich, NR4 7UA

1 INTRODUCTION

Breakfast cereals are complex foods and in order to understand their textural chracteristics it is important to study their mechanical properties as they vary with composition and structure. Breakfast cereals are composed of biopolymers, principally starch and gluten. The mechanical properties of these polymers are determined to a large extent by their glass transitions[1,2]. The combination of these polymers and the addition of other constituents in the process, such as water and sugar, alters the glass transition and the final properties of the flakes. Studies of mechanical properties and the glass transition in wheat starch and amylopectin, gluten and their mixtures as a function of water content have been reviewed by Slade and Levine[3].

This paper reports the mechanical behaviour of different breakfast cereal flake materials which were milled and then moulded as bars and conditioned to different water contents. The flakes comprised a control formulation and examples where one or more of the components had been subtracted. A hot press technique was used to reconstitute the ground flakes as bar shaped specimens to remove the geometry and structure effects and allow comparison of the matrix properties. Dynamic mechanical thermal analysis together with some impact failure tests were used to compare the response of multiple component systems with that published for simpler one and two component materials which were also in the form of hot-pressed bars.

2 EXPERIMENTAL

2.1 Sample preparation

Wheatflakes were processed from wheat grains involving pressure cooking, pelletising, flaking and toasting stages. Wheatflakes were produced with different ingredients removed from or added to the control formulation (c), which comprised initially [by % weight]: wheat [87], malt [2], sodium chloride [2], and water [9]. Some other flakes were made from wheat alone (w) comprising [by % weight] wheat [91], water [9]. Other flakes were produced by adding sucrose (w+s, c+s) or fructose (c+f) at 13% level such that the wheat: sugar ratio was 6.0±0.1:1.

Samples were ground using a laboratory grinder (Type A10 IKA Labotecknik, Staufen, Germany). The initial water content was typically 2-3% (wet weight basis, w.w.b) and sufficient water was mixed with 12 g of powder to give a 20 % water content (w.w.b) prior to moulding. The pressing procedure was similar to that outlined by Kalichevsky et al.[4] and Livings[5] and is detailed more fully elsewhere[6]. Strips 24 mm long and 8 mm wide were cut off and the sides sanded to ensure they were smooth and parallel for the DMTA analysis. They were then conditioned for 2 to 3 weeks over saturated salt solutions to give a water content range 7 to 33 % (w.w.b). The higher water contents were achieved by leaving samples over water.

For the Charpy test, the same procedure was used although 20 g of material was required to give plaques 3 mm thick. Strips 13.3 mm wide and 65 mm long were cut off and the sides again smoothed with sandpaper. The test pieces were notched (width 0.4 mm and various depths) using a low speed diamond - impregnated wafering saw (Buehler Isomet, Illinois, USA).

2.2 Instrumentation

2.2.1 DMTA measurements. The Polymer Laboratories Dynamic Mechanical Thermal Analyser (DMTA) was used in the single cantilever bending mode at a frequency of 1Hz, and strain $\sqrt{2}$ (corresponding to a nominal peak to peak displacement of 23 μm). The heating rate was 2 °C min^{-1}. The glass transition (T_g) was defined from the maximum peak in tan δ.

2.2.2 Charpy test. The impact properties were obtained using a Zwick 5102 testing machine with 0.5, 1 or 2J hammers. Two samples were tested at each notch depth and four unnotched samples. The energy to break was corrected for the pendulum air resistance by subtracting the free swing energy loss.

3 RESULTS AND DISCUSSION

3.1 DMTA

3.1.1 Effect of water content. DMTA scans were performed on all samples at different water contents. Figure 1 shows the typical variation of the bending modulus, E', for the control samples (c) which were at water contents from 7 to 18 %(w.w.b.) at the start of the experiment. Similar behaviour was also observed in the other samples of different composition. With increasing water content the T_g, as indicated by the maximum in tan δ shifted towards lower temperature as observed previously for gluten and amylopectin and their mixtures[4,7-9]. This confirms the role of water having a plasticising effect on the fabricated wheatflake samples. Figure 2 shows the temperature-dependent E' and tan δ response for the c+f samples. The decrease in E' over the temperature range (-30 to 120 °C) was greater for the fructose-containing than the control sample by 2.5 compared to 2 orders of magnitude. Kalichevsky et al.[7] also observed this phenomenon in gluten/fructose (2:1) mixtures and explained this effect by sucrose behaving like a viscous liquid which reduces the gluten rubbery modulus.

A shoulder occurred after the fall in E' at the highest water content where the sample became stiffer. A similar response was reported by Kalichevsky et al.[10] for amylopectin who discussed its origin in terms of water loss during the temperature scan and an increase in

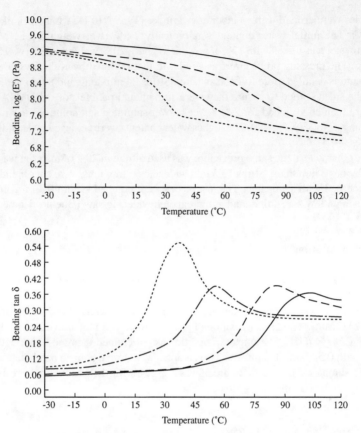

Figure 1 *DMTA Log E' and tan δ for the control samples for different water contents as a function of temperature. (———) 7%, (— —) 9%, (—·—) 14%, (······) 18%*

crystallinity in samples of 24% (w.w.b.) water. Figures 1 and 2 contrast the tan δ response of the two materials. The tan δ peak became smaller for the c+f sample and larger for the c sample with increasing water content. This may be compared with the effect of sucrose for c+s samples, for which the tan δ peak does not change significantly with moisture content[11]. The tan δ peak intensity and width respectively slightly decreased and became somewhat broader with increasing water content for w+s samples[6] compared to it becoming more intense and somewhat narrower with increasing water content for w samples[12]. Kalichevsky and Blanshard[9] and Kalichevsky et al.[10] found a more marked increase in the tan δ peak height with increasing water content in amylopectin/gluten (1:1) mixtures and amylopectin, respectively. Kalichevsky et al.[10] cited evidence from the literature for biopolymers and ionomers to support these observations. In general, they gave multiple components, crosslinking and order as reducing the size of the tan δ peak height for different proteins. The effect of sugar addition at a given water content was to broaden the tan δ peak and increase its height, as observed by Kalichevsky and Blanshard[13] for amylopectin/fructose mixtures.

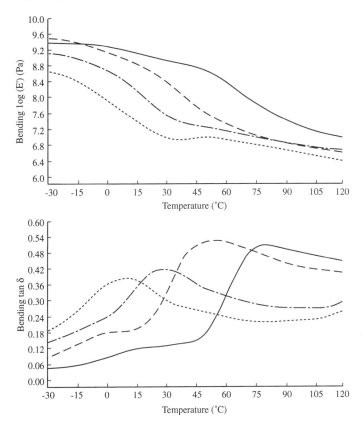

Figure 2 *DMTA Log E' and tan δ for the control + fructose samples for different water contents as a function of temperature.* (——) 9%, (— —) 13%, (——--) 17%, (------) 24%

In Figure 3, the T_g is plotted against water content for a wheat sample (w) and one having the control formulation (c). DMTA results obtained from Kalichevsky et al. for gluten[7] and amylopectin[4] were added as a comparison. In the available water content range, the T_g for the control and wheat samples were intermediate between those of amylopectin and gluten. Figure 4 shows the variation of the bending modulus with water content determined at room temperature (20°C) for the different samples. At any particular water content the bending modulus, E', was lower in the presence of sugars, although at lower water contents of 7 to 10 % (w.w.b) the modulus was approximately constant for the different samples. This is in general agreement with the changes in the room temperature Young's modulus measured by a three point bend test on gluten / fructose[7], amylopectin / fructose[14] and starch / glucose[15] mixtures.

In Figure 5, the T_g is plotted against water content. For the different compositions, the T_g fell from 100°C to 15°C with increasing water content from 7 to 22 % (w.w.b). Figures 4 and 5 show that sucrose and fructose appears to have depressed the T_g by a similar amount at the different water contents but that they affect the stiffness properties more at higher water contents. Kalichevsky et al.[7], in their studies of sugar addition to gluten (1:10),

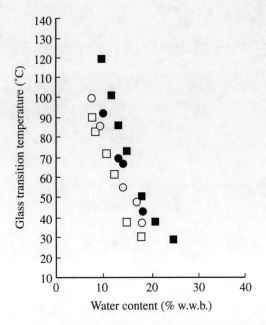

Figure 3 *T_g of amylopectin[4] (■), gluten[7] (□), wheat (●) and control (○) formulations as a function of water content*

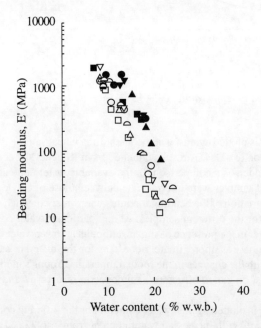

Figure 4 *E' at 20°C as a function of water content for the different compositions[6]: (■) control; (▲) wheat+salt; (●) wheat+malt; (▼) wheat; (◠) control+fructose; (□) control+sucrose; (○) control+sucrose-salt; (△) control+sucrose-malt; (▽) wheat+sucrose*

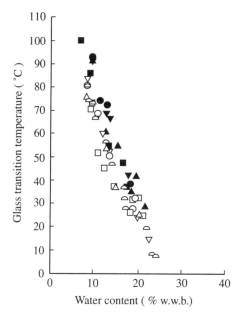

Figure 5 *T_g as a function of water content for the different compositions:*
(■) control; (▲) wheat+salt; (●) wheat+malt; (▼) wheat; (ᴄ) control+fructose;
(□) control+sucrose; (○) control+sucrose-salt; (△) control+sucrose-malt;
(▽) wheat+sucrose

commented that T_g was not greatly influenced as a function of water content in comparison with the mechanical properties.

3.2 Charpy test

An estimate of the critical energy release rate, G_c, was made based on the energy to break, W*, for different notch sizes as described by Plati and Williams [16]:

$$W*=G_c B\phi$$

where B is the sample width, D is the sample depth and ϕ is a calibration factor which depends on the ratios a/D and 2L/D where a is the crack depth and 2L is the span. Values of ϕ were calculated by Plati and Williams for this geometry and were used in the calculations reported here. Figure 6 is a plot of W* against BDϕ and the values of G_c are included in Table 1. An estimate of the fracture toughness, G_c, indicates that this was slightly higher for the sugar-free sample at low water content. The value of G_c increased dramatically from 100 to 10000 J m^{-2} corresponding to a decrease in the modulus from 1000 to 10 MPa with increasing water content as samples approached their glass transition. The energy to break unnotched bars, W*, behaved similarly (Table 1).

The present study shows that although samples were glassy at low water content and ambient temperature, the removal of sucrose or fructose made the samples harder to break or tougher. The fracture toughness has also been shown to be more sensitive than the modulus to the particle size fractions used to prepare the pressed samples[17]. In the context of sample

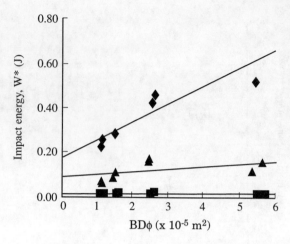

Figure 6 *Charpy test data of energy to break, W*, for the notched c+f formulation samples as a function of BDϕ at water contents; (■) 9%, (▲) 15%, (♦) 21%,)(w.w.b.), where B is the sample width, D is the sample depth and ϕ is a calibration factor* [16]

Table 1 *The energy to break, W*, and the critical stain energy release rate, G_c, determined for unnotched and notched samples, respectively, at different water contents (w.w.b)*

Water Content (% w.w.b.)	W* (J)	G_c (J m^{-2})	E' (MPa)
control (c)			
8	0.047 ± 0.002^a	214 ± 16^a	1843
13	0.047 ± 0.001^a	348 ± 16^a	553
17	0.19 ± 0.03^a	1179 ± 652^a	365
control + sucrose (c + s)			
7	0.034 ± 0.002^a	157 ± 17^a	1000
14	0.11 ± 0.02^a	1132 ± 257^a	316
19	1.35 ± 0.01^a	11890 ± 2010^a	12
control + fructose (c + f)			
9	0.027 ± 0.001^a	79 ± 20^a	1000
15	0.3 ± 0.1^a	1240 ± 808^a	253
21	0.8 ± 0.1^a	8325 ± 1256^a	16

a = standard error of mean

breakage, Kalichevsky et al.[7,14] reported that the addition of fructose (at the 1:2 level) to gluten or amylopectin reduced the water content below which their samples were brittle in three point bend tests. Kalichevsky and Blanshard[13] also reported the maximum force to break or yield was much greater for amylopectin alone than in the presence of sugars.

4 CONCLUSION

The stiffness of hot-pressed test pieces of ground wheat flakes of various composition has been compared with that reported in the literature for gluten and amylopectin and their mixtures with each other and with sugars. The glass transition of flaked wheat, and also more complex formulations of wheat, salt and malt, fell approximately within the envelope of results published for amylopectin and for gluten. The general features of the gluten and amylopectin systems were exhibited by the more complex samples, although water content affected the rubbery modulus less than previously reported. The tan δ peak was smaller at high water content for sugar - containing samples contrasting with that observed for flake material without added sugar and for wheat alone. The latter behaviour was closer to that reported for amylopectin and amylopectin/gluten (1:1) mixtures. At 20°C, addition of sucrose or fructose increasingly reduced the stiffness as the water content was increased from 7 to 22 % (w.w.b). Charpy impact tests showed that the energy to break and the toughness decreased on adding sucrose or fructose when all samples were glassy (modulus of ≥ 1 GPa). Comparison at higher water contents became more difficult because of differences in T_g, but G_c increased markedly by up to two orders of magnitude on increasing water content from 7 to 21% (w.w.b.) corresponding to a similar two orders of magnitude fall in modulus.

5 ACKNOWLEDGEMENT

This work was supported by a MAFF-DTI LINK grant under the Food Processing Sciences Programme involving Weetabix Ltd, APV Baker, Campden Food and Drink Research Association, University of Nottingham and Novo Nordisk (UK) Ltd. The authors acknowledge the scientific and financial support of the MAFF and LINK participants.

References

1. H. Levine and L. Slade, In 'Dough Rheology and Baked Product Texture', eds. H. Faridi and J.M. Faubion, Van Nostrand Reinhold, New York, 1990, p 157.
2. H. Levine and L. Slade, In 'The Glassy State in Foods' eds. J.M.V. Blanshard and P.J. Lillford, Nottingham University Press, Nottingham, 1993, p. 333.
3. L. Slade and H. Levine. In 'The Glassy State in Foods', eds. J .M. V. Blanshard and P. J. Lillford , Nottingham University Press, Nottingham, 1993, p. 35.
4. M.T. Kalichevsky, E.M. Jaroszkiewicz, S. Ablett, J.M.V. Blanshard and P.J. Lillford, *Carbohydr. Polym.*, 1992. **18**, 78.
5. S.J. Livings, 'Physical Properties of Starch Wafers', Ph.D. Thesis, University of Cambridge, 1994.
6. D. M. R. Georget and A .C. Smith. *Carbohydr. Poly.*, 1995, **28**, 305.
7. M.T. Kalichevsky, E.M. Jaroszkiewicz and J.M.V. Blanshard, *Int. J. Biol. Macromol.*,

1992, **14**, 257.

8. G.E. Attenburrow, A.P. Davies, R.H. Goodband and S.J. Ingman, *J. Cereal. Sci.*, 1992, **16**, 1.

9. M.T. Kalichevsky and J.M.V. Blanshard, *Carbohydr. Polym.*, 1992, **19**, 271.

10. M.T. Kalichevsky, J.M.V. Blanshard and R.D.L. Marsh, In 'The Glassy State in Foods' ed. J.M.V Blanshard and P.J.Lillford, Nottingham University Press, Nottingham, 1993 p. 133.

11. D. M. R. Georget and A .C. Smith, In 'Agri-Food Quality', Royal Society of Chemistry, London, 1996, In press.

12. D. M. R. Georget and A.C. Smith. *J. Therm. Analys.*, 1996, In press.

13. M.T. Kalichevsky and J.M.V. Blanshard, *Carbohydr. Polym.*, 1993, **20**, 107.

14. M.T. Kalichevsky, E.M. Jaroszkiewicz and J.M.V. Blanshard, *Polymer*, 1993, **34**, 346.

15. A.L. Ollett, R. Parker and A.C. Smith, *J. Materials Sci.*, 1991, **26**, 1351.

16. E. Plati and J.G. Williams, *Poly. Eng. Sci.*, 1975. **15**, 470.

17. D.M.R. Georget, P.A. Gunning, M.L. Parker and A.C.Smith, *J. Materials Sci.*, In press.

DEFORMATION MECHANICS OF INDIVIDUAL SWOLLEN STARCH GRANULES

L.R. Fisher[1,2], S.P. Carrington[2] and J.A. Odell[2]

[1]School of Interdisciplinary Sciences
University of the West of England
Coldharbour Lane, Bristol BS16 1QY

[2]H.H. Wills Physics Laboratory
University of Bristol
Tyndall Avenue, Bristol BS8 1TL

1 INTRODUCTION

Theories of the viscoelastic properties of starch gels are generally based on the viscoelastic properties of individual swollen starch granules.[1] To date, however, such properties of individual granules have only been poorly characterised.[2] The properties of the granule have been thought to derive from an amylopectin gel, but there are reported observations which suggest the existence of a membrane enclosing the starch gel.[3-6] We report here direct measurements of deformation as a function of applied force for individual swollen granules, and discuss the existence and contribution from the granule membrane.

2 MATERIALS AND METHODS

All measurements refer to potato starch (supplied by Sigma) in distilled water, at room temperature (25 °C) (after pre-swelling).

Deformation measurements were performed using an adaptation of a device originally designed for measuring the force required to detach a deformable particle (such as a biological cell) from a surface to which it is adhering.[7] In this apparatus, flexible micropipettes are used both to hold the particles and to apply a deforming force. The magnitude of the force is calculated from the bending of one pipette. The spring constant of the pipette is calibrated by hanging small weights at the tip and measuring the resultant deflection. The scale of the experiment is illustrated in Figure 1(A), in which a potato starch granule is held by suction under water on the tip of a micropipette. With increasing water temperature, the granule begins to swell at $T = 59$ °C (Figure 1(B, C)) and is fully swollen at 61 °C (Figure 1(D)). Note the presence of a "tongue" of material sucked into the pipette under the retaining pressure of approximately 10^5 Pa, indicating the high deformability of the swollen grain (Figure 1(D)). This point is further illustrated in Figure 1(E-H), in which a pre-swollen granule under water at 25 °C, is sucked into a micropipette. The granule, apparently undamaged by this manipulation, can subsequently be ejected from the pipette by reversal of the pressure gradient. The granule shown in Figure 1(E) is typical of the appearance of swollen starch granules of this type, in that one end of the granule is quite smooth, while the remainder is highly convoluted.

We have used three different methods to deform these swollen starch granules.

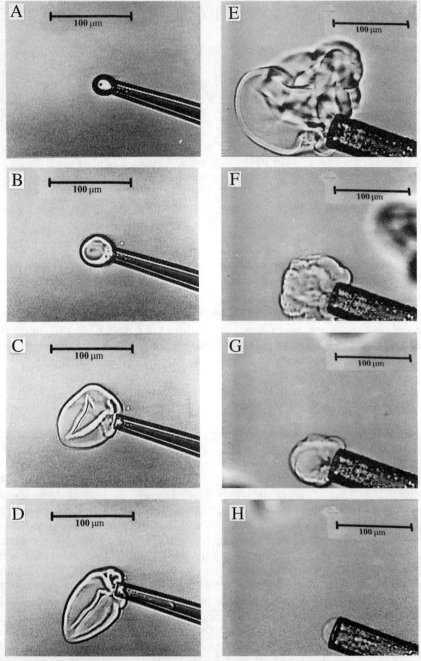

Figure 1 *(A-D): Swelling of an individual potato starch granule in water at 59-61 °C. Note the "tongue" of swollen starch being sucked into the pipette (D). (E-H): Swollen potato starch granule being sucked completely into a pipette under hydrostatic pressure.*

2.1 Compression Between Two Surfaces of Large Radius

This technique is shown schematically in Figure 2(A), with an actual videomicrograph given in Figure 2(B). A swollen granule is picked up from the bottom of a glass Petri dish using a micromanipulator controlled micropipette and light suction, and is placed against the side of an aluminium disc. A second, highly flexible micropipette, whose tip has been melted to form a sphere of radius approximately 500 µm, is used to compress the granule by different amounts. The granule deformation is measured as a function of applied force. The bending of the pipette is calculated from the difference in the distances moved from the tip (measured microscopically) and the base of the pipette (measured by position transducer).

2.2 Hydrostatic Manipulation in a Conical Capillary

Here (Figure 3(A, B)), a swollen starch granule is introduced into a capillary tube of conical internal cross-section. A hydrostatic pressure, produced by a water head mounted on a cathetometer, is used to drive the granule towards the narrower part of the capillary. Granule displacement and shape can be measured as a function of applied pressure.

2.3 Indentation

This is a combination of the first two techniques, and was adopted to give a more controlled experiment with a better defined geometry. A capillary of conical cross-section, similar to that in Figure 3, is used to hold a swollen granule in place under slight negative pressure. A flexible micropipette with a very narrow tip (ca. 6 µm diameter) is used to apply an indentation force to the granule. The tip displacement is measured as a function of applied force. This technique is shown schematically in Figure 4(A), with a videomicrograph from an actual experiment being shown in Figure 4(B). In this videomicrograph, the tip is displaced by some 70 µm from initial contact at the granule surface.

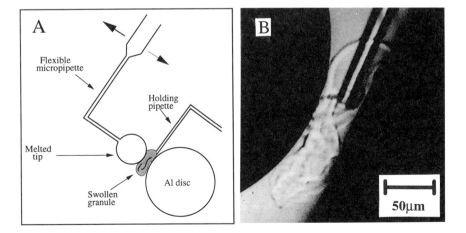

Figure 2 *(A): Schematic of the set-up for compression between two surfaces of large radius. (B): Actual videomicrograph.*

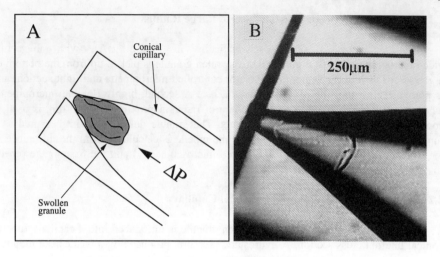

Figure 3 *(A): Schematic of the set-up for hydrostatic manipulation in a conical capillary. (B): Actual videomicrograph.*

3 RESULTS AND ANALYSIS

3.1 Compression Between Two Surfaces of Large Radius

Figure 5(A-D) shows a swollen granule under increasing compression. The orientation of the granule is such that the convoluted region of the surface is in view. The same set of folds remains under deformation, with neither original folds disappearing nor new folds being created. The deformation is entirely reversible, the granules behaving as elastic bodies. There is no measurable birefringence (between crossed polars) under deformation, which is consistent with a weak gel of low cross-link density.

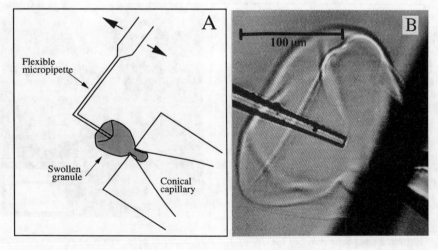

Figure 4 *(A): Schematic of the set-up for indentation. (B): Actual videomicrograph.*

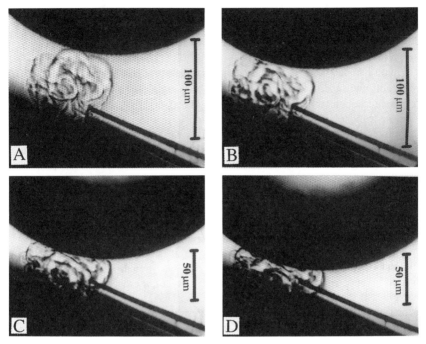

Figure 5 *(A-D): Progressive compression of a swollen starch granule.*

Figure 6 is a plot of force against displacement for three swollen starch granules compressed between two surfaces.

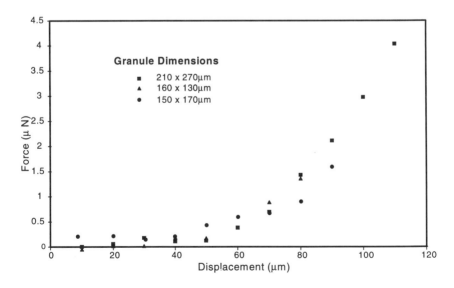

Figure 6 *Plot of force against displacement for the compression of swollen starch granules. (First granule dimension indicates the compression direction.)*

The striking feature of this plot is that the force-displacement curves appear to superimpose for different sized granules. This was generally found to be the case for the granules tested, although a granule with a more extreme aspect ratio was much softer than the typical results given here.

The data has been analysed as an elastic contact problem between three spheres of different radii and moduli.[8] The compression force is predicted to scale with the square root of granule contact radius. On this basis the data for granules of different sizes superimpose well to yield apparent moduli for the granules. The derived modulus for initial compressions is 17 Pa, increasing to 233 Pa for large compressions. This apparent stiffening of the granule may be ascribed to initial contact with the convoluted surface. As the compression proceeds the area of contact increases and the convolutions in this region progressively pull out. Deformation of the granules requires an increase in surface area so that the granule membrane would be subject to an increasing tension, this would lead to an increasing force for compression and hence an apparent increase in modulus.

3.2 Hydrostatic Manipulation in a Conical Capillary

Hydrostatic pressure actually extrudes a swollen granule from the capillary tip, demonstrating once more the extreme deformability of these granules. However, it is noticeable that if the convoluted end of a granule is placed towards the capillary tip, much higher pressures are required for extrusion, suggesting that this part of the surface is less deformable than the smooth end.

3.3 Indentation

Different granule conformations at a conical capillary tip may be produced by sucking the granule into the capillary by different amounts. Compare Figure 4(B) (approximately 10 % inside the capillary) with Figure 7(A) (approximately 80 % inside the capillary). A series of deformations under progressively increasing indentation force for the latter case is shown in Figure 7(A-D).

Figure 8 is a plot of force against displacement for the indentation of a narrow tip micropipette into three different conformations of a swollen starch granule. The "diameter" of the amount of the granule outside the capillary is also indicated in Figure 8.

This data can be analysed using two approaches from elasticity theory, which consider a similar indentation geometry. The first is the indentation of a semi-infinite elastic body by a flat punch with a circular base, and the second is a polar singular indentation of an elastic sphere.[9] Modulus (G) values can be obtained for both of these cases and the results are presented in Table 1.

Table 1 *Modulus (G) values obtained from the indentation of a starch granule at three different conformations, using two geometric approaches from elasticity theory.*

Starch "diameter" (μm)	G (Indentation of semi-infinite body) (Pa)	G (Polar indentation of a sphere) (Pa)
206	20	15
80	84	52
60	296	199

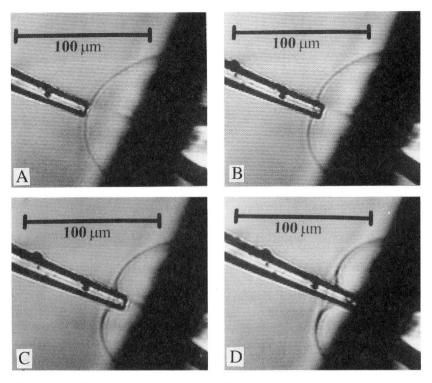

Figure 7 *(A-D): Progressive indentation of a swollen starch granule.*

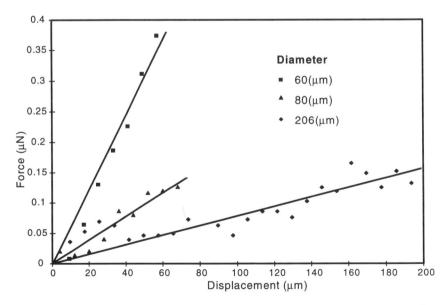

Figure 8 *Plot of force against displacement for the indentation of a swollen starch granule held at different conformations at a capillary tip.*

The results are surprisingly linear, with a significant stiffening as more of the swollen granule is sucked into the pipette. Visual observations during this process suggest that the surface membrane is being pulled tighter (compare Figure 4(B) with Figure 7(A-D)). Under extreme indentation the applied strains are complex, but would be well beyond the linear range of response for a gel model.[10] Thus the observed linearity of the force-indentation curves cannot easily be explained if the mechanical properties of the granules arise from an amylopectin gel alone. It is likely that, as with the compression experiments, there is a significant stiffening of the starch granule from a tensile contribution in a membrane. Such a model can also account for the dramatic stiffening of the granule as it is sucked into the pipette, since the stretched surface membrane supports an increased tensile force.

The direction of application of the force may also be reversed if the narrow pipette is attached to the granule by suction (Figure 9(A,B)). Note that the surface features remain despite high deformations.

3.4 Granule Inflation

On rare occasions, it has been possible to puncture a granule with a narrow pipette tip (the pipette being full of water), and then to apply hydrostatic pressure to the inside of a swollen granule. The result of one such experiment is shown in Figure 10, which gives a time sequence at t = 0 (A); t = 40 ms (B); t = 120 ms (C); t = 520 ms (D), upon application of a pressure of approximately 10^5 Pa. **These observations provide clear evidence that a membrane capable of supporting such a pressure bounds the swollen granule.** Note also that as the granule is inflated the folds tend to pull out, but only partially at the convoluted end (Figure 10(D)). If this were simply a water filled membrane, the requisite tension to support such a pressure would be of the order of 4000 mNm^{-1}; approximately 1000 times the rupture tension of a red blood cell membrane.[7]

Figure 11 shows a swollen starch granule being extruded by increasing hydrostatic pressure from the tip of a capillary with conical cross-section. In this particular case, a "bulge" was observed in the membrane surface (top left of Figure 11). This implies that the membrane is strain-softening, and the surface region of the "bulge" is subject to a relatively high local strain. The "bulge" disappears upon reduction of the pressure difference, and recurs in the same place when the pressure is reapplied.

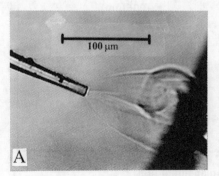
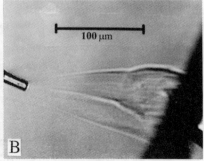

Figure 9 *(A,B): Videomicrographs showing progressive stretching of a swollen starch granule partially sucked into a conical capillary.*

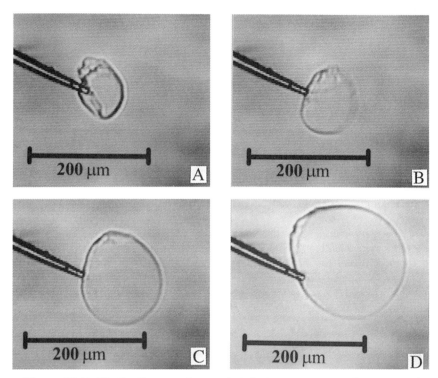

Figure 10 *(A-D): Swollen starch granule inflating at an internal pressure of ≈ 10⁵ Pa. Time: t=0 (A); t=40 ms (B); t=120 ms (C); t=520 ms (D).*

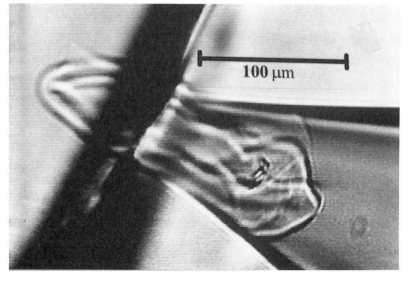

Figure 11 *Swollen starch granule exhibiting a surface "bulge" (top left) under extrusion pressure.*

4 CONCLUSIONS

We have demonstrated that it is possible to measure and model the mechanical properties of individual swollen starch granules. The granules behave as reversible and highly deformable bodies. The results of our compression and indentation mechanical tests on isolated potato starch granules point to the existence of an effective boundary membrane. The quantitative results from both techniques agree, with the effective modulus increasing as the boundary membrane tension is increased by deformation. The existence of a boundary membrane is strongly supported by our observations of the inflation of granules and the creation of surface "bulges", which we can only explain on the basis of a strain-softening membrane.

There is a very interesting dietary implication, since potato starch is one of a group of starches known to be resistant to degrading enzymes in the gut.[11,12] The membrane structure reported here could well be a contributing, or even dominating, factor in this resistance.

The mechanical tests show that there can be a considerable contribution from the tensile stress in the membrane to the apparent stiffness of the granule. This would lead to a more complex model for the interpretation of the deformation properties of individual swollen starch granules, than would arise from an amylopectin gel alone. Such a model should in turn lead to a better understanding of the viscoelastic behaviour of starch suspensions, which is of great significance for the properties and processing of food.

Acknowledgements

The authors would like to thank the following for helpful discussions during this work; Prof. John Mitchell and Lee Hartley regarding starch and Dr. Robert Arridge regarding elasticity problems. The financial support of the BBSRC and the Royal Society is gratefully acknowledged.

References

1. I.D. Evans and A. Lips, *J. Texture Studies*, 1992, **23**, 69.
2. I.D. Evans, D.R. Haisman, E.L. Elson, C. Pasternak and W.B. McConnaughey, *J. Sci. Food. Agric.*, 1986, **37**, 573.
3. K. Svegmark and A.-M. Hermansson, *Food Structure*, 1991, **10**, 117.
4. K. Svegmark and A.-M. Hermansson, *Food Structure*, 1993, **12**, 181.
5. K. Svegmark, S. Kidman and A.-M. Hermansson, *Carbohydr. Polym.*, 1993, **22**, 19.
6. J. Adler, P.M. Baldwin and C.D. Melia, *Starch/Stärke*, 1994, **46**, 252.
7. L.R. Fisher, *J. Chem. Soc. Faraday Trans.*, 1993, **89**, 2567.
8. L.D. Landau and E.M. Lifschitz, 'Theory of Elasticity', 1959, 1st Eng. Ed., 35.
9. A.I. Lur'e, '3D Problems of the Theory of Elasticity', John Wiley, 1964, 112 & 273.
10. L.R.G. Treloar, 'The Physics of Rubber Elasticity', Clarendon Press, 1975, 3rd Ed., Chapter 4.
11. H.N. Englyst and J.H. Cunnings, *Am. J. Clin. Nutr.*, 1987, **45**, 423.
12. S.G. Ring, J.M. Gee, M. Whittam, D. Orford and I.T. Johnson, *Food Chem.*, 1988, **28**, 97.

MODELLING AND EXPERIMENT ON THE FLOW OF NATIVE WHEAT STARCH GRANULES CONCENTRATED IN WATER.

J. R . Melrose[†], J. H. van Vliet[†], R. C. Ball[†], W. Frith[*] and A. Lips[*]

Cavendish Laboratory, University of Cambridge
Madingley Road, Cambridge CB3 OHE U. K.[†].
and
Unilever Research, Colworth Laboratory
Colworth House, Sharnbrook,
Bedfordshire, MK44 1LQ, U.K.[*]

1 INTRODUCTION

Native, uncooked, starch granules suspended at high concentrations in a hydrophilic medium show a complex rheology.   At low shear rates such systems shear thin with viscosity decreasing with increasing rate whilst at high shear rates they can shear thicken[1].   For a wide range of non-food applications we need to understand the key features of the granules which control this behaviour.   Two of us have recently measured the rheology of waxy maize starch granules to this end[1], here we present new data on wheat starch.   It would be highly desirable to have predictive models consistent with experiment, but there is currently only a limited capability for direct mathematical modelling of flow in concentrated particulate systems.   We therefore turn to techniques of mesoscopic computer simulations of systems of particles in fluid flow[2].   In this paper we report some preliminary work in which a simple model of swollen native granules is implemented on the computer and compared successfully with experiment.   A key feature is that the surface of the granule is modelled as an elastic but porous layer.

1.1 Rheological characterisation of native starch granules in fluids.

We have already reported the details of our experiments[1].   Starch granules were added to various solvents at room temperature, we previously reported on corn starch (waxy maize) in ethane diol and water, here we report on fractionated wheat starch granules in water.   The granules remain intact but swell by factors 1-3.   The swelling can be detected using small angle light scattering[1] but also estimated by rheological measurement.   At low dilution the viscosity is constant with respect to shear rate and its value relative to the solvent viscosity, η_r , increases linearly with volume fraction, F, according to the well known Einstein relation : $\eta_r = 1 + (5/2) \Phi$. Figure 1 shows some data for 4mM (fractionated) wheat starch granules in water; for contrast we also show similar data for granules in ethane diol.

By measuring the relative viscosity against weight fraction, W, one can find the factor , G, converting weight fraction to volume fraction by fitting to $\eta_r = 1 + (5/2) (GW)$ and inferring the granule swelling from the density of the dried granule relative to that of the solvent.   By this route we find a swelling of dried granule volume by 2.3 in water, but by a smaller factor, 1.7, in ethane diol.   It is unclear if this difference between the two solvents is just in the swelling of the core of the granule or if the granule in water has an extra corona.   The rheological method of course sidesteps the distribution of particle sizes which is available through the light scattering route.

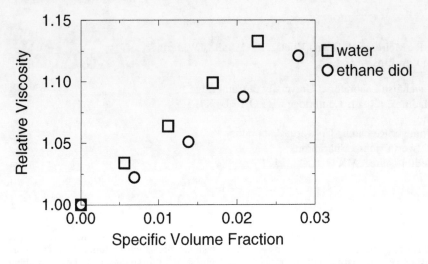

Figure 1.   *Measurements of the viscosity of low concentrations of wheat starch in water and ethane diol.*

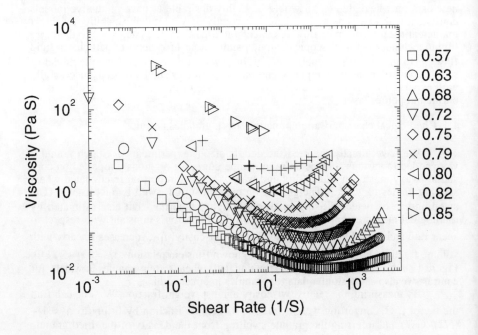

Figure 2 *Rheology data for wheat starch in water at high concentrations .*

In concentrated systems, the rheology is far from constant. Figure 2 shows data for wheat starch in water at high volume fractions. We see steep (relative to other systems) shear thinning at low shear rates and shear thickening at high shear rates. The measurements of this rheology are very delicate as care must be taken to avoid wall slip, particle migration and sedimentation problems[1]. The measurements were made using a novel stress history technique applying stress pulses by use of a constant stress cone and plate rheometer.

1.2 Simulation technique for concentrated particles in fluid.

Modelling the flow of particles concentrated in a fluid medium is a significant challenge. However, in the limit of concentrated systems the relative motion of the surfaces of close particles must dominate the viscous interactions. Indeed for purely hard spheres the fluid within narrow gaps determines a viscous force which diverges to infinity for decreasing gaps between surfaces in relative motion. We retain just these 'near field' viscous terms in our algorithm. The motion of the particles is quasi static, that is it occurs under a balance between dissipative (or viscous) forces and torques which depend on relative motions of particles, and the more usual conservative forces which are defined by a potential between particles. The equation of motion is that of Newton with the accelerations set to zero. :

$$- \mathbf{R} \cdot \mathbf{V} + \mathbf{F_C} = 0 \qquad (1)$$

The viscous force $- \mathbf{R} \cdot \mathbf{V}$, is assumed to be linear in the 6N velocity/angular velocity vector $\mathbf{V}$ and the resistance matrix $\mathbf{R}$ has coefficients which depend only on the separation of the surfaces of nearest-neighbour particles. The conservative force, $\mathbf{F_C}$, is given by direct interactions between the granules. Flow is driven by what are known as Lee-Edwards boundary conditions[3] on systems of spheres in a box surrounded by and interacting with periodic images of themselves.. Nearest neighbours are identified by the construction of Delaunay tetrahedra [4]. This guarantees that all particles with gaps below $\sqrt{2} - 1$ diameters are identified as neighbours.

Note the algorithm (1) is quite different from traditional molecular dynamics where masses and accelerations are included. To move the particles the code needs to compute the forces $\mathbf{F_C}$ and find velocities by inverting the sparse matrix $\mathbf{R}$. This leads to a highly correlated motion of the particles; for example, if a force were applied to a line of particles, in one computer time step the whole line would move.

1.3 Model for swollen native starch granules.

To proceed we must model the granules with the conservative and viscous interactions in (1). To do this we take inspiration from other systems and in particular those that show a shear thickening effect akin to that of the starch granules shown in figure 2. Dense colloidal suspensions of hard spheres with Brownian forces exhibit both thinning and thickening effects qualitatively similar but quantitatively much weaker than that of figure 2. The micro structural origins of shear thickening have long been speculated upon but are only recently being revealed through simulation techniques[5-7]. At high shear rates particles are driven ever closer together and experience ever stronger viscous interactions which lead to particles in low relative motion forming ever larger clusters. The effect is highly sensitive to surface conditions and interactions : on the one hand repulsive conservative forces will keep particles apart and limit the effect, whilst on the other hand surface features which enhance the viscous forces will enhance the effect. Indeed two of us have recently shown[7] that data on the thickening of polymer coated particles[8] can be predicted if the model accounts for the enhanced viscous force that results from fluids having to flow under the squeezing of particles through the polymer brush on their surface.

We model the surface of a swollen starch granule analogous to a polymer coat on a colloid particle. The swollen granule is modelled as a core and corona : a sphere with a deformable surface. The surface is elastic, which results in repulsive conservative forces if surfaces are deformed by pushing against one another, and it is porous to the fluid, so that surfaces which are pushed together and move relative to each other have a strong viscous force opposing their relative motion which results from the movement of the fluid through the pores. We will publish elsewhere the details of this model, but the viscous forces are derived from expressions given by others[9,10] in the context of polymer coats. The model has three dimensionless parameters such that in terms of the thermal energy k_BT and particle diameter d, the layer is of thickness L d, it has porosity L_p d and it signficantly distorts under a force $Q (k_BT/d)$.

A similar model (with similar results to those below) would result by treating the whole granule as a deformable porous sphere (in practice for a small enough L_p the coat thickness is irrelevant). However, we think it likely that the core and corona model is appropriate and that the surface is different from the bulk : either a particularly swollen amylose layer or even that the system may have absorbed proteins at its surface (although we ignore any charge effects) . There is evidence that[11] even highly swollen cooked granules have a surface membrane.

2 SIMULATION RESULTS

We performed simulations at high concentrations of particles, (where we define the volume fraction to include the core of the particle and its surface layer). Parameters were chosen so that the rheological response at the higher volume fractions is quantitatively close to that of figure 2. The volume fraction 0.63 is geometrically significant. It is known as random close packing, above this volume fraction packed spheres must either be in ordered structures (which they are not in the simulations below) or have deformable surfaces which are pushed into one another.

2.1 Data

The simulations are carried out in dimensionless units. We predict absolute values by inputting the granule size, viscosity of the base fluid (water) and temperature. Figure 3 shows the simulation rheology for a range of volume fractions with parameters L = 0.1, L_p=0.005 and Q = 10^5. These parameters were chosen to give rheology quantitatively similar to the data set of figure 2, although an explicit fitting exercise is left for future work. (The parameters are quite different, in particular Q is much larger, from those of polymer coated particles.)

3 DISCUSSION

It is only recently that we have had the ability to compute from particle models rheology in quantitative closeness to experiment. At the highest volume fractions, above close packing, there is clear agreement between figure 2 and 3 with steep shear thinning followed by steep shear thickening on increasing the shear rate. It is important to realise that without the viscous forces due to the porous surface (i.e. a simulation with just the elastic component at the surface) the model would not have shown this steep shear thickening. The simulations clearly indicate the importance of close packing as above this volume fraction (0.63) the surfaces must contact strongly and the shear thinning is far steeper. Below close packing, the simulation is far flatter in the shear thinning region. One reason why the starch data may not show this is that the granules have a weak attractive potential between themselves or even a charge at their surface which is not included in the model. We are examining these possibilities at present by both theory and experiment.

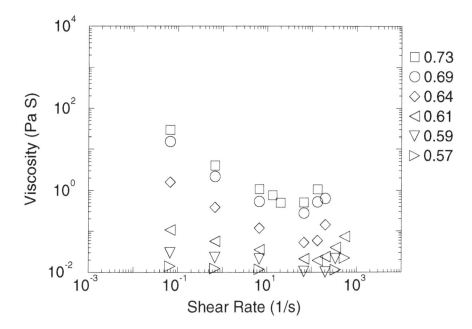

Figure 3 *Rheology data from the simulation model.*

The simulations give insight into the changes determining rheological effects. In the shear thinning region the dominant contribution to the stress is from the conservative/elastic part of the surface forces; the thinning response itself is due to the shear induced change of the angular distribution around a given granule of its contacting neighbours. At the onset of shear thickening, however, the elastic forces begin to be overcome by the applied stress, the surfaces are pushed into deeper contact and the dominant contribution to the stress comes from the viscous forces. Viscous forces resist relative motions of the particles. Therefore the strong viscous interaction between pushed together granules leads to clusters of particles moving as relatively rigid units. within the bulk. For the clusters to lead to a rise in viscosity they must be elongated[2,5,6]. We are developing a new theory of this flow induced clustering[12]. Monodisperse spheres at high concentrations can order under flow and in this case shear thickening occurs with a loss of order. However, shear thickening is not in principle an order disorder effect : in polydisperse systems without ordered flow (as is likely the case for the starch granules) shear thickening still occurs.

More generally, we hope that the quantitative capability of the simulation tool in conjunction with experiments will lead to a better understanding of the role of granule surfaces in rheology and future control of rheology through manipulation of the surface condition and variability between starch varieties.

Acknowledgements

We thank A. Donald, T Waigh, L. Silbert and R. Farr for useful discussions, C. Kousah for help in preparation of the manuscript. The Cavendish team thank the DTI/colloid technology project co-funded by the DTI, Unilever, Schlumberger and ICI, the EPSRC Colloid hydrodynamics grand challenge and the BBSRC for funding.

References

1.   W. J. Frith and A. Lips, *Adv. in Colloid and Intf. Sci.*, 1995 **61**, 161.
2.   J. F. Brady and G. Bossis, *Ann. Rev. Fluid Mech*., 1988 **20**, 111 ; J. R. Melrose and R. C. Ball, *Adv. in Colloid and Intf. Sci.*, 1995 **61**, 19.
3.   A. W. Lees and S. F. Edward, *J. Phys.*, 1972, **C5**, 1921.
4.   R. C. Ball and J. R. Melrose, unpublished 1996.
5.   J. F. Brady, T. N. Phung, and G. Bossis, *J. Rheol.*, 1995 **313**, 181.
6.   J. R. Melrose and R. C. Ball, *Europhysics Lett.*, 1995 **32**, 535
7.   J. R. Melrose and R. C. Ball, submitted *Phys. Rev. Lett.*, 1996.
8.   W. J. Frith, P. d'Haene, R. Buscall and J. Mewis, in press *J. Rheol.*, **40**(*4*), 531 1996.
9.   G. H Fredrickson and P. Pincus, *Langmuir,* 1991 **7** , 786. A. A. Potanin and W. B. Russel, *Phys. Rev. E,* 1995 **52**, 730.
10.   L. R. Fisher, in *Proc. of Starch: structure and function* 1996 *ibid.*
12.   R. Farr, J. R. Melrose and R. C. Ball unpublished 1996

ANALYTICAL DEVELOPMENTS: MOLECULAR AND MICROSTRUCTURAL CHARACTERIZATION

S. Hizukuri, Y. Takeda, J. Abe, I. Hanashiro, G. Matsunobu, and H. Kiyota

Faculty of Agriculture, Dept. of Biochemical Science and Technology
Kagoshima University
Korimoto-1, Kagoshima 890, JAPAN

1 INTRODUCTION

Starch is solid solar energy produced by photosynthesis in plants. It is abundantly found in the storage organs of plants, such as seeds and roots. It is produced in a reasonably pure state by a simple process from these sources and is easily convertible to many substances by chemical and biochemical process. It is expected to be a potential, renewable resource to play important roles in various fields of human life in the next century.

Starch is composed from only glucose, looks a simple material and yet its primary structure is not fully understood. The functional properties of starch differ from one source to another. For examples, pasting or gelatinization behaviors as revealed by amylograph or differential scanning calorimetry. These behaviors are exhibited based on the integrated and cooperative functions of their molecules and the molecular organization of the granules. However, molecular structures are considered to be the most basic factors because they affect the higher structures of starch granules.

Starches are comprised of two major molecules, amylose and amylopectin. Amylose is further subclassified into linear and slightly branched molecules. In addition, an intermediate material between amylose and amylopectin has been found in the amylopectin fraction of high amylose starches, such as amylomaize and wrinkled pea[1].

We now report on the up-to-date structures of amylose and amylopectin and the transformation of crystalline structures and functional properties of the starch granules through disorganization and recrystalization.

2 STRUCTURES OF AMYLOSE

Amylose is defined as a linear molecule but it has been well recognized that some molecules are slightly branched by $(1{\rightarrow}6)$-α-linkages. The branched amylose appears to have a similar function to make helical-inclusion complexes with iodine or other complexing agents such as 1-butanol and lipids. Furthermore, the β-limit dextrins of the branched amylose also shows the close properties to the original amylose, and thus it has different properties from amylopectin. No effective methods for the separation of linear and branched molecules are known. However, the structures of both linear and branched molecules can be characterized in some detail as follows.

The average number of the branch linkage ($\overline{NB}$) of branched amylose can be determined by the assay of the non-reducing residues of the β-limit dextrin of amylose. The molar fractions of linear and branched molecules are calculated from the numbers of the branch linkages of whole amylose and the β-limit dextrin according to Eq. (1)[2,3], or by the direct measurement of the amount of branched molecules by tritiation of reducing residue[4].

$$\text{Branched molecules (\% by mol)} = \frac{\text{No. of branch linkages of amylose}}{\text{No. of branch linkages of } \beta\text{ - LD}} \times 100 \quad (1)$$

The molecular size of branched molecules could be estimated from the molecular size of the β-limit dextrin by assuming the β-amylolysis limits as 40%. The value was found from the relationship between the β-amylolysis limit and the amount of branched molecules[1]. From these values of branched and whole amyloses, the molecular size of the linear fraction of the amylose can be calculated. Table 1 shows the molecular characteristics of amyloses from various origins. The molecular size of the branched molecules are 1.5- to 3.0- folds of those of linear molecules. The average molecular sizes of cereal amyloses are much smaller than those of tuber and rhizome starches. The amounts of the branched molecules are mainly in the range of 25-55% by mole. The average number of branch linkages of the branched molecules were in the wide range of 4-18 which we have investigated so far, but the percentage of branch linkages was in the narrow range of 0.27-0.68%. It implies that the branched molecules are a well-defined molecular species irrespective of their origins. The 40% of β-amylolysis limit suggests that the branch linkages are frequently located rather near the reducing terminal and/or they have multiple branched side chains.

The fine structures of branched amylose were analyzed through β,i,β-degradation[5]. For example, the branched properties of a barley amylose ($\overline{DP}n$ 1340) having 3.3 branch linkages on average and containing 27% (by mole) of branched molecules are discussed below[6]. The β,i,β-degradation products of the amylose were separated into two fractions, soluble small molecules and large molecules complexed with 1-butanol. The small and large molecules were 74% and 26%, respectively, by mole. The $\overline{DP}n$ of the small molecules were 17.4 and contained 1.3 G_2 residues, which implies 2 A chains per Ba-chain of original amylose and the average span length ($\overline{SL}$) between branch linkages was found to be 6.0 or slightly longer than the value on average. The mixture of branched dextrins had 1 to 2.5 or 3.0 maltosyl side chains. The $\overline{SL}$ between two branch linkages or the length from the reducing end of the molecules were suggested to be 1 to 60 by DP of the pullulanase-debranched dextrins. The $\overline{DP}n$ of the large molecules were 1070 and linked 1.7 G_2 residues, which corresponded to 3 A

Table 1 *Structural Properties of Some Amyloses*

Amylose	Branched molecule (%)	$\overline{DP}$ n			B / L	Branch linkage			
		Whole	Branched (B)	Linear (L)		Whole (n)	Whole (%)	Branched (n)	Branched (%)
Cereals									
Wheat (9[a])	39±0.10[b]	1230±229	1950±446	760±230	2.6	4.5±0.73	0.38±0.13	12.5±4.42	0.68±0.29
Rice (7)	36±7.6	1010±70	1410±154	810±70	1.7	2.4±0.84	0.23±0.08	6.5±1.53	0.46±0.09
Amylomaize (3)	44±2.2	710±20	1040±59	450±39	2.3	2.0±0.1	0.28±0.02	4.5±0.49	0.43±0.07
Maize	48	960	1320	630	2.1	2.1	0.22	4.4	0.33
Average	42	980	1430	660	2.2	2.8	0.28	7.0	0.48
Others									
Kuzu	53	1460	1950	910	2.1	3.7	0.25	6.8	0.35
Lily	39	2300	2780	2000	1.4	3.9	0.17	10.0	0.36
Tapioca	42	2660	3280	2210	1.5	6.8	0.21	16.1	0.49
Sago (H[c])	62	5090	6820	2260	3.0	10.4	0.25	18.3	0.27
(L[c])	41	2490	3050	2100	1.5	6.8	0.27	16.4	0.54
Average	47	2800	3580	1900	1.9	6.3	0.23	13.5	0.40
Overall average	45	1990	2620	1350	2.0	4.7	0.25	10.6	0.43

a. Number of varieties or classes; b. SD; c. High and low viscosity

chains per Ba chain of original amylose and the $\overline{SL}$ was 360. These results suggest that branched amylose has some tiny clusters of short chains.

3 STRUCTURE OF AMYLOPECTIN

Amylopectin is a branched polysaccharide constructed from hundreds of short $(1\rightarrow4)$-α-glucan chains, which are interlinked by $(1\rightarrow6)$-α-linkages. These chains of amylopectin are classified into A, B, and C chain. The A chains are non-branched, the B chains are branched at the C-6 position(s), and the C chain is the only one having a reducing residue. The ratio of the A to B chains are useful for characterization of the mode of branching of the amylopectin. However the reported values vary among the investigators. This appears to be due to the accumulated errors of the experimental data and the methods employed[1]. The values from 1.0 to 1.5 to 1.0 seems to be reasonable. The values are consistent with the cluster and Meyer's structures but not with those of Haworth and Staudinger. Since as early as 1930, several structures have been proposed for amylopectin by several investigators. However, nowadays the concept of cluster structure proposed by Nikuni and French has been accepted and several lines of evidence for the structure have been accumulated[1]. We have quantitatively analyzed the cluster structure by gel-exclusion chromatography and sequential enzymic analyses, namely, β,i,β-degradation[5] and γ,i-degradation[7]. These analyses give detailed information on the mode of branching.

The debranching enzymes, isoamylase and pullulanase, specifically hydrolyzed the branch linkages and produce the short linear chains. The size distributions of these chains have been characterized by gel-exclusion chromatography or high performance anion-exchange chromatography. The latter technique with pulsed amperometric detection has an advantage over the former because it is possible to separate of individual chains with high sensitivity, although quantitative assay can be done by a limited range up to DP 17[8].

In order to analyze the chain-length distribution in greater detail, the technique has been improved by using two columns in tandem and selection of suitable elution conditions, and then the neutral chains of potato amylopectin were separated on a base line up to 80 or slightly above[9]. Under these conditions, the relative chain-length (CL) distributions (by % peak area) of up to CL 60 were compared on eleven amylopectins from different species and the characteristics of the distributions were visualized by the difference of each corresponding peak-intensity. The relative areas of each chain of arrowhead amylopectin were tentatively selected as standards, because the $\overline{CL}$ of the amylopectin was the mid value of these specimens. The difference exhibited periodic waves of DP 12, except edible canna and yam amylopectins, where the period might be DP 15. Therefore, the chains were fractionated into *fa*, DP 6-12, *fb₁*, DP 13-24, *fb₂*, 25-36 , *fbₙ*, DP>37 and the amounts of these fractions are listed in Table 2. The $\overline{CL}$ of amylopectin appeared to be dependent on the amount of the *fa* fraction. This may suggest that a cluster length of most amylopectin is CL 24. The crystalline types of starch granules appear to be dependent on the amount of the *fa* fraction. The large and small amounts of the fraction decrease and increase the $\overline{CL}$ of amylopectin and produce A and B type structures, respectively.

The relative amounts of these fractions would be determined by the relative activities of the isoenzymes of starch debranching enzymes with different specificities and starch synthases or the involvement of debranching enzyme.

Table 2 *Differences in Chain-Length Distributions by Source[a]*

Source	DPn of isoamylolyzate	Crystalline type	Relative area (%)				fa / fb_{1+2+n}
			fa	*fb₁*	*fb₂*	*fbₙ*	
Waxy rice	18.9	A	29	50	11	9	0.41
Wheat	20.1	A	27	49	14	10	0.38
Rice	21.6	A	27	52	12	9	0.38
Barley	19.3	A	27	50	14	9	0.36
Arrowhead	20.5	Ca	25	53	13	9	0.33
Sweet potato	22.1	A	21	47	17	15	0.27
Normal maize	22.0	Ca	24	54	13	9	0.31
Lotus	22.6	Cc	23	55	13	9	0.30
Yam	23.2	B	18	56	15	11	0.22
Potato	23.6	B	18	48	15	18	0.22
Edible canna	25.7	B	20	56	15	9	0.25

a. Fractions were as follows: *fa*, DP 6-12; *fb₁*, 13-24; *fb₂*, 25-36; *fbₙ*, >37.

4 TRANSFORMATION OF CRYSTALLINE STRUCTURES OF STARCH GRANULES

Starch granules are microcrystalline in nature, showing characteristic X-ray diffraction patterns called A, B, and C types depending on the plant source and environmental temperature[1]. The C types are probably mixtures of the A and B types in various proportions. The crystalline domains are constructed by left-handed double helices, which are arranged in a parallel manner[10]. The double helices are packed densely in A type but there is room for ample water molecules at the center of the hexagonal packing in B type. We have investigated the cause of the formation of these crystalline structures for many years and have found that the basic factor appears to be the average chain-length from the relationship between the $\overline{CL}$ of amylopectins and the crystalline structures and from the crystallization of amylodextrin with varied CL[11]. The modifiers of the crystallization have been suggested to be the environmental temperature[12] and the microenviroments of chloroplasts or amyloplasts containing various materials. The

The $\overline{\text{CL}}$ shorter than 20 glucosyl lengths and those longer than 22 lengths produced A and B types, respectively, as mentioned above[11]. The lengths between 20 and 22 produced A, B, and C types depending on the conditions of modifiers. The high and low temperatures are advantageous for A and B types, respectively. Lipids and some ions with high lyotropic numbers shifted the crystallization toward A type[13]. At present, only sulfate ion, among the many materials tested, induced B type[14]. The B to A type transformation is possible by heat-moisture treatment. Our procedures are possible from A to B type transformation.

The first step of the transformation is to unwind the double helix and to disorder its arrangement. The second step is to restore the double helix and rearrange the helix to the ordered state. A key is to carry out these processes in the granular state, because it is unable to make a starch granule *in vitro*. The first step was performed by staining the granule with I_2-KI solution, which transforms the double helix of the A, B, or C structure to the single V helices complexed with iodine. The second step was carried out by removal of iodine from the complexes with thiosulfate solution. When the granules had been completely destroyed the double helix by iodine, it was gelatinized immediately after the removal of iodine. However, when it was performed in a solution containing high concentrations of ammonium or aluminum sulfate, they maintained granular structures and restored both birefringence and A, B, or C type X-ray diffraction depending on the temperature of the treatment although the intensities of the diffraction lines were reduced.

When normal maize starch was subjected to the above treatments, it produced B type below 10 °C, C type at 20 °C, and A types above 35 °C. Thus, it was confirmed that environmental temperature was one of the factors affecting the crystalline polymorphs as originally reported on soybean seedings[11]. In the case of sweet potato, it produced B type below 20 °C, C type at 30 °C, and A type at 35 °C or above. Thus, the transition temperatures for X-ray types were higher than those of normal maize starch. This is consistent with the idea that the large amounts of the *fa* fraction induce A type. Concomitantly, the pasting properties of these starches were also altered.

The enthalpy values of the gelatinization were reduced to about two thirds of the original value, due to the lower crystallinity as suggested by X-ray diffraction. There was a clear trend that the gelatinization temperature increased with the elevation of the second step. The peak temperatures of DSC of the recrystallized normal maize starches removed iodine at 10, 20 and 35 °C were 43, 45 and 50 °C, respectively. These values were much lower than that (66 °C) of original maize starch. Thus, it reduced the gelatinization temperature greatly. The enthalpy values decreased, also suggesting lower crystallinity.

The pasting properties characterized by an RVA were also altered by the transformation. The pasting temperature was decreased, maximum viscosity was increased and breakdown was increased with a decrease of the temperature of the second step (Table 3).

Table 3 *Pasting Properties of Recrystallized Maize Starches at Varied Temperatures*

Crystal. temp. (°C)	X-Ray type	T_O (°C)	T_M (°C)	T_C (°C)	ΔH J/g	Relative	Pasting temp. (°C)	Max. viscosity (RVU)	Breakdown (RVU)	Setback at 40 °C (RVU)
				DSC				RVA		
Native	A	62.4	66.3	83.8	15.3	100	74.9	286	115	189
10	B	35.2	42.6	78.3	9.1	60	55.0	339	267	209
20	C	36.3	45.0	77.8	8.9	58	56.8	350	266	198
40	A	37.4	50.2	81.7	8.7	57	59.0	372	310	131

These results suggest that the A-, B-, and C-type structures are determined primarily by the CL of amylopectin and are influenced by the environmental temperature and that the pasting properties are also considerably influenced by the environmental temperature. This new process of the modification of granular structures of starch is of great interest for understanding the mechanism of the formaton and the structure-function relationships of starch granules.

5 CONCLUSION

The molecular structures of starch components have been well analyzed over the last 10 or 20 years, yet we do not fully understand the details of these individual starches or the granule. Much investigation is still needed on this old, mysterious material.

Genetic studies to produce new starches with unique functional properties by creating various transformants with modified starch-metabolizing enzymes have great potential for the utilization of starch. The detailed analysis of starch molecules provides valuable knowledge for these studies.

REFERENCES

1. S. Hizukuri, 'Carbohydrates in Food' (Ed. A.-C. Eliasson), Marcel Dekker, New York, 1996, 347.
2. S. Hizukuri, Y. Takeda, and S. Imamura, *Nippon Nogei Kagaku Kaishi*, 1972, **46**, 119.
3. Y. Takeda, S. Hizukuri, C. Takeda, and A. Suzuki, *Carbohydr. Res.*, 1987, **165**, 139.
4. Y. Takeda, N. Maruta, and S. Hizukuri, *Carbohydr. Res.*, 1992, **227**, 113.
5. S. Hizukuri and Y. Maehara, *Carbohydr. Res.*, 1990, **206**, 145.
6. S. Hizukuri, Y. Takeda, and H. Kiyota, Manuscript in preparation.
7. S. Hizukuri and J. Abe, 'Plant Polymeric Carbohydrates' (Ed. F. Meuser, D. J. Manners, and W. Seibel), Royal Society Chemistry, Cambridge, 1993, 16.

8.   K. Koizumi, M. Fukuda, and S. Hizukuri, *J. Chromatogr.*, 1991, **585**, 233.
9.   I. Hanashiro, J. Abe, and S. Hizukuri, *Carbohydr. Res.*, 1996, **283**, 151.
10. A. Imberty, A. Buléon, V. Tran, and S. Pérez, *Starch/Stärke*, 375, **43**, 1991.
11. S. Hizukuri, T. Kaneko, and Y. Takeda, *Biochim. Biophys, Acta*, 1983, **760**, 188.
12. S. Hizukuri, M. Fujii, and Z. Nikuni, *Nature*, 1961, **192**, 239.
13. S. Hizukuri, Y. Takeda, S. Usami, and Y. Takase, *Cabohydr. Res.*, 1980, **83**, 193.
14. S. Hizukuri, M. Fujii, and Z. Nikuni, *Biochim. Biophys. Acta*, 1960, **40**, 346.

STUDIES ON THE AMYLOLYTIC BREAKDOWN OF DAMAGED STARCH IN CEREAL AND NON-CEREAL FLOURS USING HIGH PERFORMANCE LIQUID CHROMATOGRAPHY AND SCANNING ELECTRON MICROSCOPY

P.M. Mathias[+], K. Bailey[+], J.J. McEvoy[+], M Cuffe[*], A. Savage[*], and A. Allen[*]

[+]Department of Biological Sciences, Dublin Institute of Technology, Kevin Street, Dublin 8
[*]Department of Chemistry, University College, Galway
[*]Department of Science, Regional Technical College, Tallaght, Dublin, Republic of Ireland

1. INTRODUCTION

During the milling of wheat grains a proportion of the starch granules in the flour is damaged by fracturing of hydrogen bonds between, and covalent bonds within, starch molecules[1]. Since damaged starch granules absorb more water than undamaged granules, flour millers control damaged starch levels in the reduction stages of milling, thereby controlling water absorption to specification required by the bread maker. Damaged starch is also the substrate for amylases which provide fermentable carbohydrate that is utilised by yeast in the long fermentation bread making processes. This affects the rheology and gassing power of dough, as well as the crumb texture and crust colour of the bread[1,2]. Thus, accurate and reproducible measurement of the degree of starch damage is important for the flour miller and the bread maker.

The main methods for the determination of damaged starch rely on the use of amylases which rapidly hydrolyse damaged starch to reducing sugars (expressed as maltose). These are then determined by titration after reduction of a standardised solution of potassium ferricyanide[3,4]. Gibson et al.[5] published an improved enzymatic method which uses fungal alpha-amylase to digest the damaged starch. The oligosaccharides are then hydrolysed by amyloglucosidase to glucose, which is quantified with a glucose oxidase/peroxidase reagent.

There are two important variables for these methods. First, the type of amylase used will influence the nature of the hydrolytic products.[5] Alpha-amylases produce glucose, maltose and alpha-limit dextrins, whilst beta-amylases have a more restricted ability to digest starch. They will only cleave to within two to three D-glycosyl residues of alpha-1,6 glycosidic bonds in amylopectin, producing mainly maltose and beta-limit dextrins[6]. In a previous study the addition of beta-amylase to alpha-amylase was shown to increase apparent starch damage values[5].

Second, there is an observed biphasic relationship between starch damage and incubation time with amylase with an initial rapid rise in starch damage values followed by a slower steadier increase. This could be due either to increased enzymic access to damaged sites with time or hydrolysis of undamaged granules. In addition, within individual granules damaged sites may exist that differ in their susceptibility to amylolytic hydrolysis[5,7]. The aim of the current study was to investigate these observations by examining the nature of the hydrolytic products of digestion of damage starch granule by two different types of amylases through the application of High Performance Liquid Chromatography (HPLC) and scanning electron microscopy (SEM). HPLC, linked with refractive index (RI) detection, provides a method for the direct qualitative and quantitative evaluation of simple sugars after incubation of flours with solutions of alpha-amylase[8]. However, greater sensitivity and the measurement of a wider range of reducing sugars can be obtained with the newer method of high performance anion exchange chromatography using pulsed amperometric detection (HPAEPAD). Carbohydrates are weakly acidic and by this method are separated at a high pH using an anion exchange stationary phase. This paper describes the use of both HPLC methods in the investigation of the amylolytic breakdown of damaged starch in both cereal and non-cereal flours.

2. EXPERIMENTAL

2.1 Materials

A range of high protein bread making flours and one low protein biscuit flour were obtained from Odlum Group Ltd., Dublin. A pregelatinised wheat flour was obtained from Flahavan's Ltd., Dublin. A damaged starch standard sample was obtained from the Campden and Chorleywood Food Research Association (CCFRA) with a value of 6.3% $\pm$ 1%. Non wheat and non cereal flours, were purchased locally.
Sugars, used to make up standard solutions for calibration purposes were obtained from Sigma Chemical Corporation.

2.1.1 Enzyme Assays

Fungal alpha-amylase (Sigma A-0273, 40 U /mg solid, from *Aspergillus oryzae*) porcine pancreatic alpha-amylase (Sigma A-3176, 29U alpha amylase /mg solid, 3.6U beta-amylase/ mg solid) and amyloglucosidase (Sigma A-7255 20800U/g solid) were obtained from the Sigma Chemical Company. They were prepared in solutions of 0.05M sodium acetate buffer pH 5.5. Glucose was determined using glucose oxidase peroxidase (GOPOD) chromogen kits obtained from Boehringer Mannheim following the manufacturer's instructions.

2.1.2 HPLC -Dionex - HPAEPAD

A Dionex DX500 (Dionex Corp., Sunnyvale, CA) system was used. Separation was by means of a Carbopac PA-1 anion exchange column attached to a GP40 gradient pump. Samples were eluted using a gradient of 0-600nM NaOAc in 100mM NaOH over 30 minutes at a flow rate of 1 ml/min. The column was reequilibrated using 100mM NaOH for 10 minutes before successive injections. Detection was by Pulsed Amperometry (ED40 electrochemical detector), and subsequent data analysis made use of Peaknet software (Dionex Corp., Sunnyvale, CA) on a Gateway 2000 PC.

2.1.3 HPLC - Refractive Index (RI)

A Shimadzu LC-10AT solvent delivery module with a Shimadzu RID-6A refractive index detector and a Shimadzu CBM-10A Communications Bus Module were used with a 25cm x 4.6mm Spherisorb, Phasesep S-10 amino column, maintained at 30°C. Chromatogram traces were recorded using Shimadzu LC Workstation Class LC-10 software operating on a Dell personal computer. The solvent system was acetonitrile/water 75:25 v/v with a flow rate of 1.5ml/min . Samples were injected on to the column through a Rheodyne 7125 valve with a fixed volume 20µl loop.

3.   METHODS

3.1.Colorimetric procedure for starch damage (Modified AACC 70-31[1])

Flour samples (100mg) were weighed into 10 ml test tubes. 2.25ml of either pancreatic alpha-amylase solution (100U/ml) or fungal alpha-amylase (100U/ml) were added to each tube and mixed. The samples were incubated at 40°C for 10 minutes and 5.4mls 0.2% v/v sulfuric acid added to terminate the enzymic reaction, followed by centrifugation at 4000 rpm for 10 minutes. 0.5 mls of filtrate was removed and added to 3.5mls 0.05M sodium acetate buffer pH5.5. 1.5mls amyloglucosidase solution (10U/ml) were added, the tubes mixed and then incubated at 40°C for 10 minutes. The glucose content was measured using the GOPOD chromogen kit. For the time course experiments samples were removed at 10 minute intervals up to 60 minutes in the initial incubation step with the alpha-amylases, followed by incubation with amyloglucosidase and glucose estimation.

3.1.1. Calculations:

mg glucose/100 ml of sample = absorption of sample x 9.1 (9.1mg /100ml, glucose standard) / absorption of glucose standard.

mg glucose / 5.5 ml sample = mg glucose / 100ml sample /100 x 5.5

% damaged starch = mg glucose/ 5.5ml sample - blank sample (containing enzyme only) x 0.9 (162/180 - factor to convert glucose to starch) x100 / B (calculated sample weight)

B = sample weight (100mg) / 7.65 x 0.5 (sample dilution)

3.2 HPAEPAD (Dionex) - Comparison of hydrolytic products from pancreatic and fungal alpha-amylases. Time course experiments

Flour samples (2g) were weighed in duplicate and mixed with 40 mls of 0.05M sodium acetate buffer pH 5.5. 5mls of either pancreatic alpha-amylase (200U/ml) or fungal alpha-amylase (200U/ml) were added to each sample and incubated at 30°C in a shaking water bath. 5 mls aliquots were removed at 10 minutes intervals up to 60 minutes, and added to 12 mls 0.2% v/v sulfuric acid to terminate the enzymic reaction. Samples were filtered and 10µl injected onto the column. Sugars were quantified by peak height ratios using rhamnose as an internal standard.

3.3 HPLC - R I. Analysis of hydrolytic products using fungal alpha-amylase.

Flour samples (500mg) were weighed into 10 ml testtubes in duplicate. 2mls of fungal alpha-amylase solution (200U/ml) were added, the tubes and incubated at 40°C for 10 minutes and then centrifuged. The supernatant was transferred to a second tube and placed in a boiling water bath for 90 seconds to destroy the enzymic activity. The tubes were removed, cooled to room temperature and centrifuged. 20ml of supernatant were injected onto the column. Sugars were quantified by comparison to appropriate external standards.

3.4. SEM - Maize and potato flour; time course experiments

Flour samples (3g) were mixed with 60mls of sodium acetate buffer pH 5.5 . 5mls of pancreatic alpha-amylase (500U/ml) or were added and incubated in a 30°C shaking water bath. At 10 minute intervals up to 60 minutes 3ml of were removed and added to 120 mls 66% ethanol solution to terminate the enzymic activity. The samples were allowed to dry in a fume hood and reconstituted in approximately 1 ml of deionised water. Up to six drops were placed on a SEM disc and allowed to dry in air. The discs were coated with gold using a 5400 Biorad Sputter for 70 seconds and examined under an Hitachi S 2400 Scanning Electron Microscope.

4.   RESULTS

4.1 HPAEPAD (Dionex)

4.1.1 Wheat flours

Figure 1 shows a typical chromatogram for the HPAEPAD method obtained for a hydrolysate of starch from CCFRA standard flour at 20 minutes incubation time with pancreatic alpha-amylase. Maltose was the main hydrolytic product, with smaller amounts of glucose and maltotriose present. Figures 2 and 3 compare the profiles of sugars after incubation of the two flours with pancreatic and fungal alpha-amylases. Figures 4 and 5 show the effect of incubation time with pancreatic and fungal amylases respectively on maltose hydrolysed from damaged starch in the two wheat flours. For both enzymes there appears to be a rapid increase in maltose up to 20 minutes and a further steady increase in maltose up to 60 minutes incubation time.

4.1.2 Maize and potato flours

Detailed amylolytic experiments were carried out on maize and potato flours. Figures 6 and 7 show the comparative effects of the two enzymes on the hydrolytic products at 20 minutes incubation time. Whereas for the fungal alpha-amylase maltose accounts for over 80% of the sugars released, incubation with pancreatic alpha-amylase appears to produce appreciable amounts of glucose and maltotriose. The ratio of glucose:maltose:maltotriose was approximately 20:50:30 for both maize and potato flours.

Experiments were carried out with the pancreatic alpha-amylase to further elucidate the action of this enzyme on maize and potato starch. Figure 8 shows the comparative amounts of sugars produced from damaged starch from both flours, demonstrating a much lower digestion of the potato starch granule. Figures 9 and 10 shows the effect of incubation time up to 60 minutes on the production of glucose, maltose and maltotriose from maize and potato starch respectively. They show an initial rapid increase of all sugars up to 10 minutes, followed by a steady increase up to 60 minutes, most notably for maltose, for both types of flour.

4.3 SEM

The relationship between this observed increase in sugars with incubation time with pancreatic alpha-amylase and physical changes in the starch granule was investigated by SEM. Figure 11 shows micrographs of the changes in the maize starch granule, and Figure 12 those in the potato starch granule up to a period of 60 minutes. The maize starch granules clearly show appreciable endocorrosion by 60 minutes, as do the potato starch granules.

4.4. HPLC - RI

Following incubation with fungal alpha-amylase for 10 minutes, amylolytic products were quantified for a range of bread making flours and other cereal and non cereal flours by HPLC - RI. The results are shown in Table 1. Maltose accounted for 68-91% of the total sugars released in the wheat flours and maize, but under 50% of sugars from barley, with glucose and maltotriose the other breakdown products. There was a significant positive correlation ($r^2 = 0.92$, $p<0.001$) between the maltose values quantified by HPLC and percentage starch damage determined by the colorimetric method, for all flours. Based on the maltose values calculations were made for starch damage in these flours using the simple calculation, mg maltose (HPLC) /mg flour x 100. These values were compared with those obtained by the modified AACC 70-31 colorimetric assay. The results, in Table 1 show good agreement between the two assay systems.

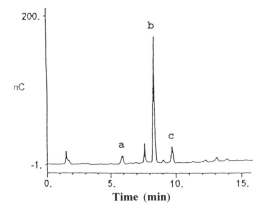

FIGURE 1. *HPAEPAD (Dionex) of a starch hydrolysate from CCFRA standard flour incubated with pancreatic alpha-amylase for 20 minutes. Eluent = gradient of 0-600mM NaOAc in 100nM NaOH over 30 minutes, flow rate, 1ml/ml. Column, Carbopac PA-1 anion exchange column. a = glucose, b= maltose, c = maltotriose.*

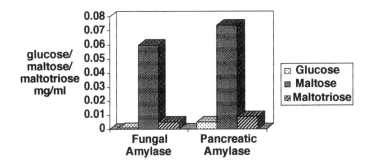

FIGURE 2. *HPAEPAD sugar profiles following incubation of CCFRA standard flour with fungal or pancreatic alpha-amylases.*

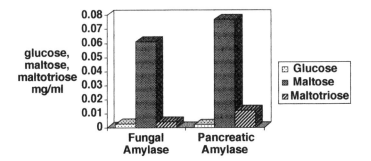

FIGURE 3. *HPAEPAD sugar profiles following incubation of high protein flour with fungal or pancreatic alpha-amylases.*

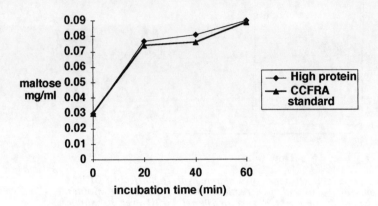

FIGURE 4. *Effect of incubation time with pancreatic alpha-amylase on maltose released from wheat flours, determined by HPAEPAD (Dionex).*

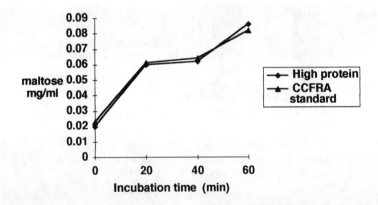

FIGURE 5. *Effect of incubation time with fungal alpha-amylase on maltose released from wheat flours, determined by HPAEPAD (Dionex).*

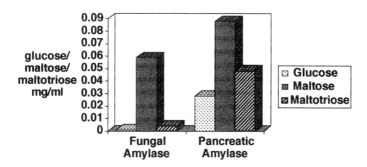

FIGURE 6. *HPAEPAD sugar profiles following incubation of maize flour with fungal or pancreatic alpha-amylases*

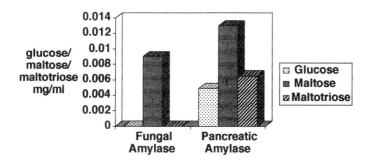

FIGURE 7. *HPAEPAD sugar profiles from following incubation of potato flour with fungal or pancreatic alpha amylases.*

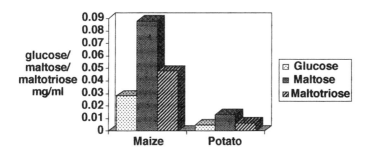

FIGURE 8. *Comparison of HPAEPAD sugar profiles following incubation of maize and potato flour with fungal or pancreatic alpha-amylases*

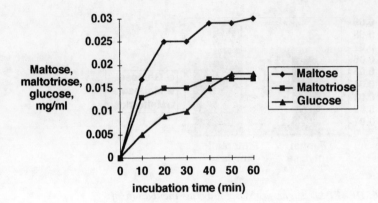

FIGURE 9. *Effect of incubation time with pancreatic alpha-amylase on levels of sugars released from maize flour, determined by HPAEPAD (Dionex).*

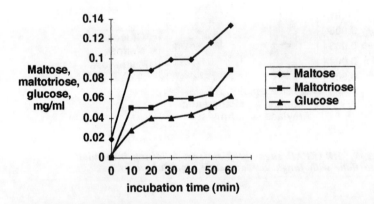

FIGURE 10. *Effect of incubation time with pancreatic alpha-amylase on levels of sugars released from potato flour determined by HPAEPAD (Dionex).*

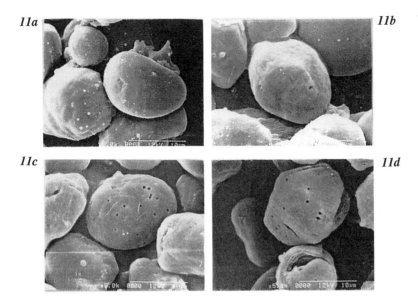

FIGURE 11. *Scanning electron micrographs of damaged starch granules taken during incubation of maize flour with pancreatic alpha-amylase. Incubation times: (a) 0 min (b) 20 min (c) 40 min (d) 60 min*

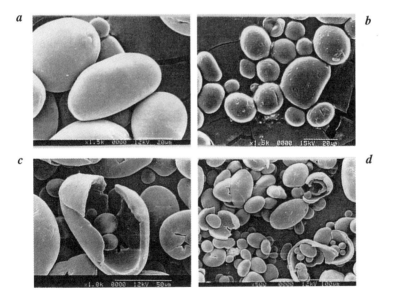

FIGURE 12. *Scanning electron micrographs of damaged starch granules taken during incubation of potato flour with pancreatic alpha-amylase. Incubation times: (a) 0 min (b) 20 min (c) 40 min (d) 60 min*

TABLE 1. Sugar concentrations obtained by HPLC RI and calculation of starch damage values expressed as mg maltose/mg flour x 100, comparison with starch damage values obtained by modified AACC 70-31 colorimetric assay

Flour	Glucose (mg/ml)	Maltose (mg/ml)	Maltotriose (mg/ml)	Starch Damage (%)	
				HPLC-RI (mg maltose/mg flour) x100	AACC 70-31
CCFRA Standard	1.2 (6)[†]	17.3 (86)	1.7 (8)	6.8	6.5
*High Protein	0.9 (5)	16.6 (91)	0.8 (4)	6.5	6.0
*Quality	5.7 (24)	16.4 (68)	2.0 (8)	6.5	6.5
*Constancy	5.5 (23)	16.5 (69)	1.8 (8)	6.5	6.3
*Superbaker	5.5 (23)	16.5 (70)	1.5 (6)	6.6	6.3
Pregelatinised	1.7 (5)	28.6 (82)	4.7 (13)	11.3	14.0
Biscuit	0.2 (2)	10.7 (91)	0.8 (7)	3.9	3.7
Maize	1.5 (9)	14.4 (84)	1.3 (8)	5.7	6.2
Barley	8.0 (43)	8.9 (48)	1.6 (9)	3.6	4.5
Potato	0 (0)	2.0 (100)	0 (0)	0.6	1.3

* Wheat flours supplied by Odlum Group Ltd., Dublin

[†]Figures in parentheses are sugars expressed as % of total (glucose + maltose + maltotriose).

5. DISCUSSION

An accurate and meaningful interpretation of starch damage values in flours is dependent on (a) the degree to which all damaged sites with individual granules are hydrolysed and (b) the extent of hydrolysis of undamaged starch granules. These factors are influenced by both the type of the amylase used in the starch damage assay and incubation time with the enzyme[5]. In this study the investigation of these factors was carried out using two types of alpha-amylase. The first, the fungal alpha amylase from *Aspergillus oryzae* which, being of reasonable purity, is the most commonly use in modern starch damage assays[1,5,10]. The second, porcine pancreatic alpha-amylase is not a pure alpha-amylase, but contains 10% beta-amylase activity, which has a more restrictive action on starch to that of pure alpha-amylase[6].

5.1 HPAEPAD (Dionex)

The application of HPLC techniques, especially the modern and sensitive HPAEPAD (Dionex) chromatography allows direct qualitative and quantitative evaluation of the products of amylolysis.

5.1.1 Wheat flours

The results for HPAEPAD in Figs. 2 and 3 suggest different amounts and different patterns of sugars were released from damaged starch in wheat flours from the two types of alpha-amylases used in this study. Incubation of wheat flours with fungal alpha-amylase produced 90% or more of maltose with maltotriose accounting for less than 10%, whereas the pancreatic amylase caused the release of higher quantities of maltotriose (10 -15% of total sugars) and to a smaller extent glucose, as well as increased overall total sugars compared with the fungal enzyme.

Increasing incubation time with both enzymes showed an increase in maltose production up to 60 minutes (Figures 4 and 5). This could be interpreted as a type of biphasic effect, with an initial rapid increase up to 20 minutes, and another increase from 40 to 60 minutes.

5.1.2. Maize and Potato Flours

These effects were seen most strikingly with maize and potato flours. Here maltotriose and glucose account for 50% of sugars released by pancreatic alpha-amylase with overall sugar production again being higher than that from the fungal enzyme (Figs. 6 and 7). Maize and potato starch have different crystalline patterns. Maize, a cereal starch has the same basic crystalline structure as wheat starches, i.e. A-type, whereas potato starch is mainly B-type crystalline structure, which is more resistant to amylolysis due to a higher amylopectin content[9] .This accounts for the much lower production of total sugars from potato flour than from maize flour during amylolysis (Fig. 8). Time course incubation experiments with the pancreatic alpha-amylase showed a rise in production of all sugars from both maize and potato flours up to 60 minutes, with maltose and maltotriose showing the sharpest increases (Figs. 9 and 10).

5.2. SEM

Micrographs of maize and potato granules taken over this incubation time demonstrate that at 20 minutes there is little endocorrosion of starch granules, whereas from 40 to 60 minutes significant endocorrosion and breakdown of the starch granules has occurred (Figures 11 and 12). Patterns of damage are seen to differ for both types of starch granule. For maize starch granule tiny holes start to appear at 20 minutes, indicative of the start of endocorrosion. These become larger with longer incubation time, when at 60 minutes granules have started to crack giving a clear picture of the crystalline structure within the granule, which shows concentric rings of ordered amylopectin while the amorphous region is being digested between the rings (Fig. 11). The potato granule with its smoother egg-shape appearance suffers amylolytic damage in a different fashion. Small cracks first appear on the surface of the granule (exocorrosion) at 20 minutes, with endocorrosion developing with longer incubation periods shown by a breaking open and hollowing out of the granule. At this point there seems to be smaller "egg" shapes granules resting within the larger broken granules. (Figure 12) This appears to be a unique observation, which warrants further investigation.

These results clarify some previous observations relating to starch damage assays. First, Gibson *et al.* [5] demonstrated a synergistic effect of the addition of a beta-amylase to alpha-amylase on the starch damage value in flour, using a colorimetric assay. Results from the present study, where pancreatic alpha-amylase contained 10% beta-amylase activity, suggests that this synergistic effect is a result of not only the release of more maltose from the damaged starch granule, but also an increase of other sugars especially maltotriose. The effect is most apparent with maize and potato flours. A possible explanation for this observation is that although beta-amylase has a more restricted ability to digest starch its action may expose a larger number of sites for digestion by alpha-amylase. The reason why this synergistic effect is more enhanced in maize and potato starch is unclear, and has not been reported previously. Furthermore, results from the present study suggest that pure alpha-amylase (in the form of the fungal enzyme) may be insufficient to hydrolyse all possible damaged starch sites. This would result in an underestimation of actual starch damage in flour. It is interesting to note that the original starch damage method of Farrand[3] used enzymes from malted barley which contain a mixture of alpha and beta amylases. This may be a reason why the Farrand method is found to give higher values for starch damage than more modern methods[10].

Second, a number of authors have reported that there is no end-point to starch damage assay, as apparent starch damage values increase with increasing incubation time with alpha-amylase[5,7]. This was confirmed in the present study, which demonstrated that for wheat flours there was an increase in maltose released, and for maize and potato flours all hydrolytic products, glucose, maltose and maltotriose increased with incubation time. SEM results suggest that up to 20 minutes incubation with amylase starch granules are relatively intact and that only damaged starch is broken down. After this endocorrosion of the granule and subsequent amylolysis of the starch contained therein accounts for the further increase in sugar production. This indicates that in routine starch damage assays incubation times of under 20 minutes are required for an accurate assessment of damage starch values.

5.3 HPLC - RI

There are surprisingly few reports on the use of HPLC with RI detection in the actual estimation of starch damage[8,10]. In the present study sugars in a range of hydrolysates from a variety of bread making and non bread making flours were quantified, following incubation with pure fungal alpha-amylase. Since maltose is the sugar of interest when considering potential fermentative and hence rheological properties of the processed flour, these are the values of interest in starch damage calculations. Results for maltose in high protein bread making flours were similar to those obtained for damaged starch standards using similar conditions by Sutton and Mouat [8] who obtained between 11.9 - 15.3 mg/ml maltose. In the present study all the high protein/bread making flours gave remarkable consistent values for maltose between 16.4 to 17.3 mg/ml (Table 1).

The diverse variety of flours used in these experiments gave a wide range of maltose values. This allowed the correlation with values from the standard colorimetric AACC method to be made with confidence. The significant correlation (r^2 = 0.92, p<0.001) between HPLC maltose values and AACC starch damage values allows prediction of starch damage from maltose values, as suggested previously by Sutton and Mouat [8]. However, in the present study values for starch damage in flours directly calculated from HPLC maltose figures were shown to correspond well with those obtained by the AACC method.

In conclusion, these studies show that HPLC with RI detection can be used with confidence and consistency to assess maltose values from damaged starch in bread making flours. These figures can then be used to directly calculate starch damage values. Therefore, this method would be of significant importance to the flour making and bread making industries for the rapid evaluation of the potential fermentative and rheological properties of the flour.

References

1. A.D. Evers and D.J. Stevens, "Starch Damage" in Advances in Cereal Science and Technology (Y. Pomeranz, Ed.),American Association of Cereal Chemists, St.Paul Minnesota, 1985, **vol . 7,** p 321.
2. K.H. Tipples, Bakers Digest, 1969, **43,** 28
3. E.A. Farrand, Cereal Chem., 1964, **41,** 98
4. AACC Method 76-30 A .Approved Methods of the AACC, 1984, American Association of Cereal Chemists, St. Paul, Minnesota
5. T.S. Gibson, H. Al Qalla and B.V. McCleary, J. Cereal Sci., 1992, **15,** 15.
6. P. Colonna, V. Leloup and A. Buleon, Eur. J. Clin. Nutr., 1992,**46,** suppl.,S17
7. W. C. Barnes, Starch,1978, **30** 114
8. K.H. Sutton and C.L. Mouat J., Cereal Sci., 1990, **11,** 235.
9. D.J. Gallant, B. Bouchet, A.Buleon and S. Perez, Eur. J. Clin. Nutr.,1992, **46,** suppl., S3.
10. J. Karkalas, R.F. Tester, and W.R. Morrison, J. Cereal Sci., 1992, **16,** 237

AMYLOLYSIS OF STARCH GRANULES AND α-GLUCAN CRYSTALLITES

V. Planchot, P. Colonna, A. Buleon and D. Gallant

Institut National de la Recherche Agronomique,
BP 1627, F - 44316 NANTES Cedex 03, France.

1 INTRODUCTION

The notion of starch degradation is an important consideration in human and animal nutrition, and the problem of resistant starch has often been investigated. Yet little is actually known about the amylolysis of starchy foods in the human gut. α–amylase (α–1,4 glucan-4-glucanohydrolase, EC 3.2.1.1.) from the pancreas is the main enzyme involved in the hydrolysis of α–1,4 bonds of starch (amylose, amylopectin) and maltodextrins. This enzyme is able to bypass α–1,6 branching points but does not cleave them. In contrast to hydrolysis of amylose, which is randomly transformed into malto-oligosaccharides, the action of α–amylase on amylopectin is a non-random process, leading to the production of maltose and maltotriose and branched α–limit dextrins composed of all the initial α–1,6 bonds and the adjacent α–1,4 linkages. Finally, when the granule is solubilised, hydrolysis of normal starches yields 70% malto-oligosaccharides (40% maltose, 25% maltotriose and 5% dextrins of higher polymerisation degree) and 30% α–limit dextrins[1].

In vivo study of hydrolysis products from pancreatic α–amylase is almost impossible since the hydrolysis products diffuse immediately towards the brush border where they are hydrolysed into glucose by the different oligosaccharidases present. The high levels of maltase-amyloglucosidase and saccharase-α–dextrinase activities *in vivo* prevent maltose accumulation, which would slow down or stop α–amylase action. However, this is not the case for *in vitro* experimental data resulting from end-point tests which can be considered as measurements of overall starch hydrolysis under conditions of enzyme limitation .

When any dispersed starch is fully hydrated, it is completely hydrolysed *in vivo*. This type of material is generally found in mashed, freshly-cooked starchy foods. However, starch usually occurs in foods in complex solid structures whose features have an influence on the hydrolysis pattern[2]. Native starches are classified as number 2 in the resistant starch classification. They form semicrystalline structures which can be considered as an amorphous matrix in which more resistant regions are embedded. Short-range molecular ordering in native granules is usually attributed to clusters of short chains of amylopectin in association with amylose chains. The main changes in starch ultra- and microstructures are the increased surface area-to-starch ratio in the solid phase, modifications of crystallinity through the effects of gelatinisation, complexing and gelation, and the depolymerisation of amylose and amylopectin. Amylose and amylopectin gels (third in the resistant starch classification), which are often found in starchy food, are interconnected three-dimensional networks formed by an interchain association with B-type crystallinity. The fraction of total crystalline material is considered to be an important factor in defining the rate and extent of enzymatic hydrolysis.

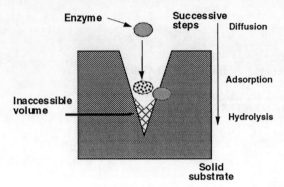

Figure 1 *Three successive steps in enzymatic hydrolysis of an insoluble material.*

In a mechanistic description, the amylolysis of solid substrate (Figure 1) involves three steps which limit the kinetics and extent of hydrolysis: diffusion of the enzyme molecule, adsorption of enzymes onto solid substrates and the catalytic event. Only the first and third steps will be discussed here.

2 DIFFUSION AND ACCESSIBILITY

Amorphous starchy substrates immersed in water swell when the water is absorbed inside the matrix, resulting in rapid hydrolysis. However, semicrystalline materials are less water-sensitive since crystallites act as crosslinks, preventing swelling. Diffusion of enzyme molecules inside the substrate is then determined by the porosity of the solid substrate and the diffusion coefficient of the enzyme inside the pores. It is difficult to probe enzyme diffusion inside starchy substrates as degradation, occurring simultaneously to diffusion, increases consequently porosity.

Starch granules and gels[3] are porous substrates with both external and internal surface area. The external surface area is determined by the size and shape of the particle. The internal surface area, which is defined by the capillary structure of the starch particles and the size of the penetrating reagent, can be in the range of 10 to 100 $m^2 g^{-1}$. Internal surface area can be measured by the solute exclusion technique using a hydrated sample: measurement is performed with the substrate in an aqueous environment identical to that in which enzymatic hydrolysis occurs. This sort of approach has been used successfully for cellulosic materials in which enzymatic degradation can be studied in the same way as for starchy materials[4].

Accessibility was studied by diffusion until equilibrium, using probe size and macromolecule concentration as the main parameters. Figure 2 shows estimates of the accessible volume of each amylose gel, according to the amylose concentration, with respect to specific surface area versus the molecular reagent diameter, as obtained by a column solute exclusion study. Accessibility is a reasonably continuous function of hydrodynamic radius. For probes in which $R_H < 10$Å, such as H^+, the volume of solvent trapped within the gel network is completely accessible to the probe. The accessibility for a protein probe, BSA, of same hydrodynamic radius (3.61 nm) than α–amylase, can therefore be described by the ratio Ce/Co, where Co would represent the probe concentration if the volume of the solvent trapped inside the material was totally inaccessible and Ce the measured probe concentration in the material. The accessibility ranges from 1 (totally accessible structure) to 0 (totally inaccessible structure). Plots of accessibility versus the amylose concentration are shown Figure 3 for BSA. The gel accessibility shows a sharp decrease when amylose gel concentration is increased from 2 to 10%. Unfortunately no equivalent expriment was conducted on native starch granule. As the fully hydrated density (starch plus water) is about 1.3, with a water uptake of roughly 50 % (dry basis), starch granules can assimilated to 200 % gel. Using this previous linear regresion, accessibility was extrapolated to nul value for a polymer concentration of 180 %. This result is in agreement with the conclusion of French[5], stating that penetration within the starch granule is limited to solutes less than about 1000 daltons. These observations could be extended to granules during biosynthesis: starch growth and amylopectin branching are two events occurring at the periphery of the granule as diffusion of both enzymes and DP 15 and 45 chains are impossible inside a starch granule.

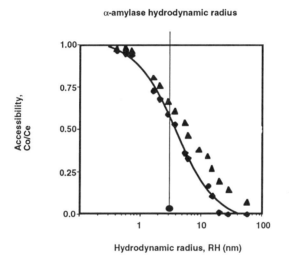

Figure 2 *Accessible volumes of 4% (s) and 8% (u) amylose gels.*

The second factor concerns the rate of diffusion of enzyme molecules inside pores. Water is entrapped in these three-dimensional networks in which macromolecular probes (enzymes) can diffuse. The diffusion coefficient (D) of bovine serum albumin in amylose and amylopectin gelled networks decreases with increasing polysaccharide concentration[6]. D decreased from 5.5×10^{-11} m^2.s^{-1} to 1.5×10^{-11} m^2.s^{-1} over the concentration range 5-15% w/w of gelling polymer. Hydrodynamic screening is considered to be a more appropriate description of the physical process responsible for diffusion delay. The relationship between the average mesh size ξ of the semi-dilute polymer solution and the hydrodynamic radius R_H of the probe particle is important.

When $R_H >> \xi$, the matrix appears as a continuum, and the Stokes-Einstein relationship may be used to describe the diffusion process, with the viscosity term being the macroscopic viscosity of the polymer solution.

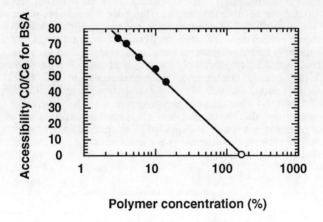

Polymer concentration (%)

Figure 3 *Accessibility versus the amylose concentration for BSA measured in gels (l) and extrapolated in starch granules (m).*

3. NATIVE GRANULAR STARCHES

3.1 Relative efficiencies of amylases from different sources

Biochemical analysis provided a mean hydrolysis value for the whole population of starch granules. Three α–amylases from *Bacillus subtilis*, *A. fumigatus* and porcine pancreas were used. Regardless of the α–amylase used, starches could be classified into two groups: one comprising those degraded more than 50% under the conditions used (1.34 nKat / mg of substrate), including waxy and normal maize, smooth and wrinkled pea and wheat; and the other comprising potato and high-amylose maize starches, for which extents of degradation were far lower than 50%. However, when an enzyme from *Bacillus subtilis* or porcine pancreas was used, easily degradable starches belonging to the first group became completely degraded when enzymatic activity was increased; whereas potato and high-amylose starches were solubilised by a maximum of 15 and 20% respectively. In our conditions, the final extent of hydrolysis was 28 and 48% respectively for potato and high-amylose starches with the enzyme from *A. fumigatus*. Due to its efficiency on granules less susceptible to enzymatic degradation, this enzyme was chosen for microscopy observations.

3.2 Microscopy observations

The action pattern of amylases on granular starches was clarified by scanning (SEM) and transmission electron microscopy (TEM), which are very powerful tools for demonstrating that the botanical origin of starch plays a major role in determining the hydrolysis pattern.

Susceptibility to amylolysis can be classified according to the intensity and manner in which granules are eroded and corroded. Most granules are first hydrolysed superficially, with the exception of high-amylose and potato starches which are very resistant to amylolysis.

Normal maize starches are hydrolysed by random deep pitting, with canals enlarging through the granules at each shell level. With α–amylase from *A. fumigatus*[7] (1.34 nKat/mg of starch), one hour of hydrolysis (15% of degradation) was enough to produce different states of degradation. Moreover, advanced degradation was observed for all intermediate shapes. Before the appearance of pores, the starch granule surface was furrowed randomly in weakened areas. Subsequently, the pores penetrated through the outer shells of the granules. When they enlarged, individual pores produced external grooves and internal corrosion channels. TEM examination showed that hydrolysis occurred mainly in the more amorphous zones (darker shells), obliterating the original "blocklet" structure observed and defined by Gallant *et al.*.[8] Final hydrolysis led to the formation of individual "blocklets" delineated by simultaneous tangential and radial hydrolysis.

For waxy maize starch, deep pitting was amplified by extensive tangential corrosion, which cut out the soft part of the shell. After one hour of hydrolysis (22% degradation) with α–amylase from *A. fumigatus*, starch granules appeared to be highly degraded in SEM. As for normal maize, all intermediate shapes of degradation were observed. No preferential hydrolysis zones were detected by TEM, demonstrating a uniform susceptibility to amylolysis.

Wheat and barley starch granules showed susceptibility zones which became pitted during subsequent endocorrosion. The pits enlarged and endocorrosion canals were visible within granules. Numerous canals then formed randomly around the granules.

Potato and most B-type starches (according to Katz's classification[9]) were slighty eroded by exocorrosion. After 4-h hydrolysis (1.6% degradation) with α–amylase from *A. fumigatus*, starch granule degradation was herterogeneous, with most granules remaining intact or being only slightly modified. The granule surface, which was smooth in the native state, roughened as soon as amylolysis started. A tiny fraction, representing less than one percent, was composed of highly degraded granules. Similar observations were made after 29-h hydrolysis (12% degradation), at which time the differences between intact and modified granules were even more obvious. Intact granules were still present, representing the dominant population even after extensive degradation. A minor population of damaged granules always consisted of only the outer granule parts. The internal ultrastructure of the potato starch granule was studied by TEM after PATAg treatment. Endoerosion occurred in the amorphous zones (darker shells) of the granule. As with SEM, no intermediate forms of degradation between native and highly damaged starch granules were detected by TEM.

Similarly, high amylose maize starch granules showed only small pores on the surface, whereas the starch granule itself was empty. Two modes of attack were observed: minor exoerosion observed by SEM and major endocorrosion observed only on thin sections by TEM. When hydrolysis was total, only the residual outer shell remained (1 to 3 μm thick).

The classical paradigm is thus that the hard and supposedly the most crystalline parts of the granules are less hydrolysed than the soft parts (Fig. 4). However, this is in contradiction with the presence in these zones of α–(1,6) linkages which should be resistant to amylolysis.

3.3 Basis for susceptibility to amylolytic enzymes

Studies based on SEM and TEM observations and the kinetics and efficiencies of different amylases enabled us to distinguish two starch families.

In the first family composed of normal and waxy maize starches, the entire starch granule population was degraded uniformly regardless of degradation extent. Even for low extents of hydrolysis (15-25%), all starch granules appeared highly degraded. The emergence of a layered internal structure in these granules suggested the existence of superimposed layers of high and low susceptibilities related to different crystallinity levels.

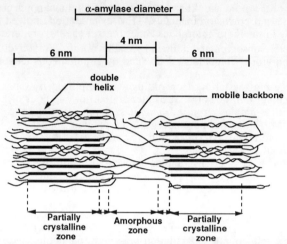

Figure 4 *Schematic representation of the arrangement of amylose and amylopectin molecules within crystalline and amorphous zones inside a starch granule, with indication of α–amylase size.*

The second family consisted of the less susceptible starches, *i.e.* potato and high-amylose maize. In this case, all granules were not equally degraded at a given time. Although most showed no or very little pitting, a few were highly degraded, with fragmenting of external shells. No intermediate state of degradation was noted. In contrast to the first group, starch granules from high-amylose maize in the first stage of degradation appeared almost intact when viewed in SEM. The degradation pattern of these resistant starches appears to take place in two steps. The first corresponds to uniform adsorption of enzyme molecules, creating superficial microporosity. Concomitantly, when the enzyme encounters a less organised structure, degradation produces macroporosity, with deeper grooves. The central part of the granule is then eroded rapidly, whereas hydrolysis is slower for the outer shell.

Thus, there were numerous exceptions to the classical paradigm that starches with an A-type crystallinity are completely degradable, whereas those with B-type crystallites are resistant to amylolysis. The relative lack of analytical tools for examining the microstructure of remaining semicrystalline material can be overcome by the choice of specifically designed substrates.

4 *DE NOVO* CRYSTALLITES AND LINTNERS

Investigations of the amylolysis of pure crystalline models would seem to be a feasible approach for simplifying studies of the amylolysis of starch granules. The influence of crystalline features was investigated using crystalline starchy substrates such as lintners (resistant to mild acid hydrolysis) and amylose spherocrystals[10], both of which have very similar levels of crystallinity. Lintners are representative of the crystalline parts of native granules and facilitate the study of amylolysis within the granule structure. However, the tridimensional shape and size of these structures cannot be easily checked. Conversely, spherulites[11] may mimic both granule morphology and crystallinity type.

4.1 Crystalline modifications

Lintners from potato, wrinkled pea and high-amylose maize starches had a pure crystalline B-type, whereas normal and waxy maize lintnerised starches had a pure A-type, cassava and smooth pea starch lintners gave more complex diffractograms characteristic of the C-type. Native wheat starch, initially of A-type, changed to C-type after acid hydrolysis, giving X-ray diffractograms characteristic of an A-B mixture. For pure A-type, the most intense bands corresponded to Bragg angles (2θ): 9.9°, 11.2°, 15°, 17°, 18.1° and 23.3°. For wheat starch lintner, the appearance of the B-type structure was marked by an additional peak at the 5.6° (2θ) Bragg angle, a decrease in the relative intensity of the peak at 15°, predominance of the 17° reflexion instead of the classical 17-18° doublet, and a splitting of the 23° peak into two individual reflexions at 22° and 24° (2θ).

When amylolysis residues were subjected to further amylolysis, no clear increase in crystallinity level was observed, regardless of the initial crystalline type. For C-type lintners, a progression towards B-type was observed, which was more or less intense depending on botanical origin. The bands corresponding to Bragg angles (2θ) 5.6°, 15°, 17° and 22° appeared more or less intensely. Surprisingly, A-type lintners (Figure 5) after enzymatic hydrolysis gave X-ray diffractograms characteristic of the pure B-type.

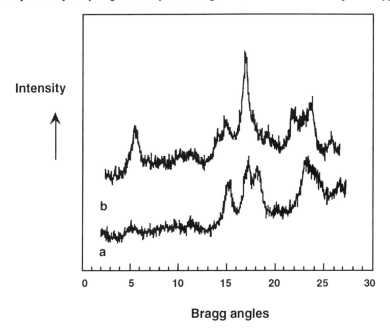

Figure 5 *X-ray diffraction patterns of lintner from waxy maize before (a) and after (b) amylolysis.*

Both A- and B-type spherocrystals showed a pure crystalline pattern, with no trace of A-type in B-type spherolites or the inverse. No transformations were observed after amylolysis.

4.2 Amylolysis extents

The main classification of studied lintners was obtained for the final extent of hydrolysis after 30 h (Figure 6), with values respectively of less than 10% for high-amylose starches, in the range of 20 to 37% for potato and C-type lintners and greater than 62% for pure A-type lintners.

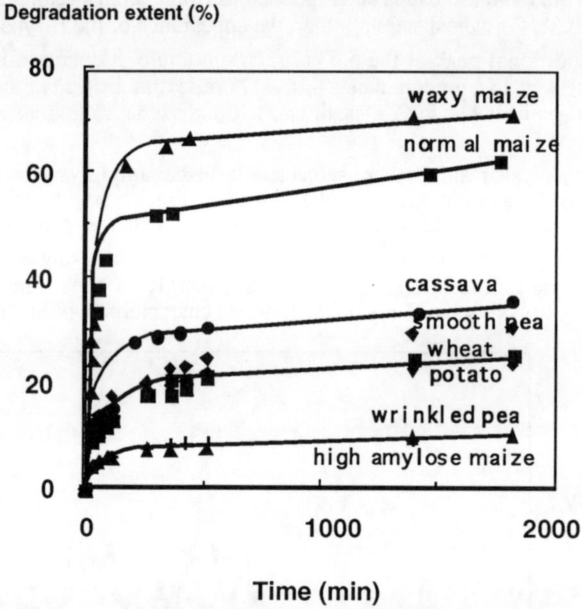

Degradation extent (%)

Figure 6 *Amylolysis kinetics of different lintnerised starches using Bacillus sp. alpha-amylase.*

The final hydrolysed fraction represented respectively 61 and 18% for A- and B-type spherocrystals (Figure 7). As for lintners, material presenting A-type crystallinity was less resistant to amylolysis than that presenting B-type.

4.3 Changes in chain-length distribution

Analysis by HPSEC showed that no sample before or after amylolysis contained a chain with a degree of polymerisation (DP) lower than 6. The identification of each DP was performed using the procedure developed for enzymatically synthesised and strictly linear α–glucans, according to Roger[12]. Results for the HPAEC and HPSEC procedures were entirely concordant for identification of the second peak composed of DP 15.

Chromatograms of A- and C-type lintners showed two major peaks corresponding to DP 15 and 25 respectively, in agreement with the description initially given by Robin *et al.*[13] The ratio of each of these peaks was in relation to the botanical origin of the starch granules. Maize and wheat lintnerised starches (Fig. 8) showed a distribution with a higher amount of DP 25 (53.7% and 57.0% respectively) in contrast to cassava and smooth pea starch lintners (34.0% and 12.0 % respectively). After amylolysis, all of the corresponding hydrolysis residues showed a distribution with a single peak (sometimes with a slight shoulder) corresponding to DP 25.

Degradation extent (%)

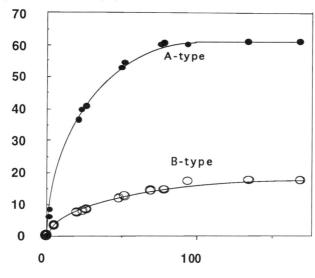

Time (h)

Figure 7 *Amylolysis kinetics of A- (l) and B-type (m) spherolites using Bacillus sp. alpha-amylase.*

RI Detector response

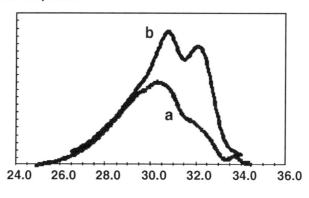

Elution time (min)

Figure 8 *HPLC-SEC chromatograms of lintnerised wheat starch before (a) and after (b) amylolysis.*

Potato starch lintner showed a distribution with a peak at DP 15 (with a slight shoulder at DP 25). The chromatograms for high amylose content in maize and wrinkled pea starch lintners showed a single broad peak with a mean average DP of around 17/18, whereas polydispersity was much greater than for the other lintners. For maize, wrinkled pea and potato lintners, the distribution pattern was unchanged after amylolysis.

5 GENERAL DISCUSSION

Analysis of the first steps in enzymatic hydrolysis enabled us to determine the heterogeneities of the organisation of outer granule shells. This should prove useful for studies of the side-by-side arrangement of amylopectin and amylose segments within the last growth ring of the native granule.

All degradation studies were conducted on a binary model in which the starch granule had the form of a highly concentrated and condensed gel, with crystallites as junction zones. There was no clear demarcation between the crystallite and amorphous phases of the starch granules. This suggests that α-glucans, though not involved in crystalline zones, are nonetheless not strictly free and mobile. As the exocorrosion of amorphous fully hydrated de novo materials cannot be simulated at the present time, reference crystalline materials offer a means of studying the susceptibility of native starch granules.

The major finding in this study was the clear relationship between the hydrolysis rate of lintnerised starches and their crystalline type. Regardless of morphology, particles with A-type crystallinity were more susceptible to amylolysis than those with B-type. A-type lintners such as those from waxy maize showed the highest rates, whereas the rate for C-type lintners (mixtures of A- and B-type structures) was dependent on the A-type ratio. Similarly, A-type spherocrystals were 3.5 times more degraded than B-type. These results confirm the higher susceptibility of A- than B-type crystals to enzymatic hydrolysis. Nevertheless, this dependence of susceptibility on crystalline type does not account for the behaviour of all studied lintners during enzymatic action. For instance, wheat starch lintner, which was composed of a mixture of A- and B-type crystals, had a rate of hydrolysis similar to that of potato starch lintner which has a pure B-type structure. As both crystalline models in this study were based on the same double-helix configuration[14,15], the chemical basis for this difference in susceptibility could relate either to an atomic level with different binding energies or to a supramolecular level with morphology and defects.

The second important observation was that polymorphic changes in A-type lintnerised starches occurred during amylolysis. These modifications are not comparable with the results of Pfannemuller[16] who found that the crystalline type of amylose precipitates was dependent on the length of amylose chains: DP 10-12 for A-type and 13 and above for B-type. In our study, the progression towards B-type structure observed during hydrolysis of both A- and C-type lintners might also have been due to the greater susceptibility of the A-type structure. During hydrolysis of C-type lintners, A-type phases might be preferentially degraded, releasing more resistant residual B-type phases. The complete conversion of A-type lintners (from waxy maize and normal maize starches) to a pure B-type structure during enzymatic hydrolysis might result from the metastability of A-type crystallites, i.e. the removal of a fragment of an A-type crystallite would cause the remaining chain to reorganise into the more stable crystalline B-type. However, constraints present in the initial crystallite could exert sufficient stress on each crystallite, thereby stabilising A-type crystallinity. Deviations from unit cell density could thus correspond to dislocations, lamellar folds and even voids. Another criterion is the minimisation of surface energy. Unfortunately, no experimental technique is currently available to assess all these structural features.

However, this dependence of susceptibility on crystalline type does not account for the behaviour of all lintners studied during hydrolysis. Knowlege of molecular chain distribution, more specifically of the ratio beween linear and single branched chains, might show that α–(1-6) linkage is also involved in amylolysis susceptibility. The exact role of residual α–(1,6) links present in A- and C-type lintnerised starches in the form of single-branched dextrins is not known. They could induce antagonistic effects such as the generation of defects in crystalline arrays or specific hindering of enzymatic action around the α–(1,6) linkage. Although the α–(1,6) zone cannot be strictly classified as a crystalline area, Buleon and Tran[17] have demonstrated, on the basis of α(1-6) linkage at the surface of a crystallite, that only two glucosyl units are needed in each strand to recover the local conformation appropriate to the crystalline unit, whether of A- or B-type. In their molecular model, all α–(1,6) and α–(1,4) linkages are included in the central channel of the double helix α under constraint. In all cases, this prevents the functioning of the classical hydrolysis acid mechanism which requires sufficient molecular mobility between glycosyl units. Therefore, an understanding of the molecular chain distribution of such substrates, and more specifically of the ratio between linear and single-branched chains, could be used to relate the presence of α–(1,6) linkages to amylolysis susceptibility. The two crystallographic models indicate that a DP 15 chain induces a crystal thickness of 5.35 and 5.20 nm in A- and B-types respectively.

Enzyme-substrate interactions need to be investigated on a molecular scale. Further studies are required to assess potential hydrolysis sites on crystalline material. It is necessary to determine the mobility and helicity of chains in the crystalline structures.

References

[1]T. Yamamoto and S. Kitahata, "Handbook of amylases and related enzymes", Amylase Research Society of Japan Ed, Pergamon Press, Oxford, 1988, p. 18.

[2]. P. Colonna, A. Buleon and V. Leloup, *Eur. J. Clin. Nutr.*, 1992, **46**, S17.

[3]. V.M. Leloup, P. Colonna and S. Ring, *J. Cereal. Sci.*, 1992, **16**, 253.

[4] R.P. Neuman and L.P.Walker, *Biotechnol. Bioeng.*, 1992, **40**, 218.

[5]D. French, "Starch: Chemistry and Technology"", R.L. Whistler, J.N. BeMiler and E.F. Paschall Ed, Academic Press Inc., London, 1984, p. 183-247.

[6]. V. Leloup, P. Colonna and S.G. Ring, *Macromolecules*, 1990, **23**, 862.

[7]. V. Planchot, P. Colonna, D.J. Gallant, B. Bouchet, *J. Cereal Sci.*, 1995, **21**, 163.

[8]. D. Gallant, B. Bouchet, A. Buleon and S. Perez, *Eur. J. of Clinical Nutr.*, 1995, **46**, 3.

[9]. J.R. Katz, Z. *Phys. Chem.*, 1930, **150**, 37.

[10]. S.G. Ring, M.J. Miles, V.J. Morris, R. Turner and P. Colonna, *Int J. Biol. Macromol.*, 1987, **9**, 158.

[11]. W. Helbert, H. Chanzy, V. Planchot, A. Buleon and P. Colonna, P., *Int. J. Biol. Macromol.*, 1993, **15**, 183.

[12]. P. Roger and P. Colonna, *Carbohydrate Polymers*, 1993, **21**, 83.

[13]. J-P. Robin, C. Mercier, F. Duprat, R. Charbonniere and A. Guilbot, *Die Stärke*, 1975, **2**, 36-45 .

[14]. A. Imberty, H. Chanzy, S. Perez, A. Buleon and V. Tran, *J. Mol. Biol.*, 1988, **201**, 365.

[15]. A. Imberty, and S. Perez, *Biopolymers*, 1988, **27**, 1205.

[16] B. Pfannemuller, *Int. J;,Biol. Macromol.*, 1987, **9**, 105.

[17]. A. Buleon and V. Tran, *Int. J. Biol. Macromol.*, 1990, **12**, 345.

SUBSTITUTION PATTERNS IN CHEMICALLY MODIFIED STARCH STUDIED BY DIRECT MASS SPECTROMETRY

P. A. M. Steeneken,[1] A. C. Tas,[2] A. J. J. Woortman,[1] and P. Sanders[1]

[1] Netherlands Institute for Carbohydrate Research TNO
Rouaanstraat 27, 9723 CC Groningen, The Netherlands

[2] TNO Nutrition and Food Research
P.O. Box 360, 3700 AJ Zeist, The Netherlands

1 INTRODUCTION

Starch is an important raw material for the production of thickeners, binding agents, adhesives, and pharmaceuticals. In order to upgrade its functional properties, starch is often modified chemically by partial substitution of the hydroxyl groups of its individual glucose units.[1] Almost all substitution reactions are carried out on more or less intact starch granules. Therefore, the question arises whether the substitution pattern is random or is affected by the semicrystalline nature of the granules.

It has to be emphasised that, in a systematic study of substitution patterns, different structural features and distance scales within the starch granule have to be taken into account.[2] Here, surface effects, reactivity of individual hydroxyl groups, of amylose and amylopectin, of amorphous and crystalline regions, and the clustering tendency of substituents along the chain can be distinguished.

In the study of these different aspects of substituent distributions, mass spectrometric methods can play a prominent role. An obvious reason for this is that substituted glucose units differ from unsubstituted ones by their mass rather than by their conformation. Another factor is the recent advent of direct mass spectrometric techniques for the study of high mass compounds.

In this contribution we address the application of direct mass spectrometry to two different problems: the monomer composition of highly substituted hydroxyethyl starch and cellulose by pyrolysis / desorption chemical ionisation mass spectrometry (Py-DCI-MS), and the clustering tendency of methyl groups along the starch chains in partially methylated starch by matrix-assisted laser desorption ionisation / time-of-flight mass spectrometry (MALDI-TOF-MS).

2 EXPERIMENTAL PART

Samples of hydroxyethyl starch (HES) and hydroxyethyl cellulose (HEC) were provided by AVEBE and Aqualon respectively. Potato starches were partially methylated in granular suspension or in solution as described previously.[2,3] The degree of molar substitution (MS) was determined with a Zeisel method.[4]

GLC-MS of HES and HEC was performed on a CP Sil-5CB capillary column with temperature programming after complete acid hydrolysis (2M trifluoroacetic acid, 125°C, 1 h) and conversion of the monomers into their perethyl hydroxyethyl glucitols.[5] Acid hydrolysates were permethylated according to Ciucanu and Kerek.[6] For Py-DCI-MS, a Finnegan MAT 8230 mass spectrometer was used with ammonia as the reagent gas.

Partially methylated starch was degraded by a sequential treatment with α-amylase from *Bacillus sp.* (Sigma) and amyloglucosidase from *Aspergillus niger* (Sigma). Hydrolysates were desalted. Preparative fractionation was performed on Biogel P2. The MS of the malto-oligosaccharides in each fraction was measured by GLC on CP Sil-5CB after conversion into partially methylated glucitol acetates.[2] For MALDI-TOF-MS, a Finnegan MAT Vision 2000 mass spectrometer equipped with an UV - nitrogen laser (λ = 337 nm, pulse duration 3 ns) was used. The matrix was 2,5-dihydroxybenzoic acid (2,5-DHB). Analyte concentration was 100 μg/ml.

3 MONOMER COMPOSITION OF HYDROXYETHYL STARCH AND CELLULOSE

Hydroxyethyl starch (HES) is prepared by reaction of starch with ethene oxide. HES differs from common starch derivatives in the sense that the substituent carries a hydroxyl group which may act as a site for additional reaction. In this way, poly(ethene oxide) chains of different length may be formed at each of the three available hydroxyl groups of a glucose unit (Scheme 1).

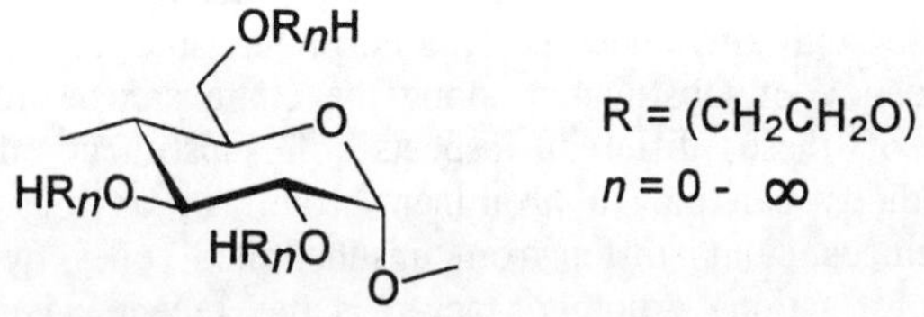

$$R = (CH_2CH_2O)$$
$$n = 0 - \infty$$

Scheme 1

Hence, the number of ethene oxide units per glucose monomer (MS) is unlimited in principle and a complex substitution pattern arises. Hydroxyethyl cellulose (HEC) is usually much more highly substituted than HES, so that HEC's were used here to illustrate the capability of our analytical methods.

3.1 GLC

The conventional method to unravel the complete monomer composition of starch and cellulose ethers involves acid hydrolysis to a monomer mixture which is analysed by GLC (-MS) after volatilisation. Usually, the monomers in the mixture

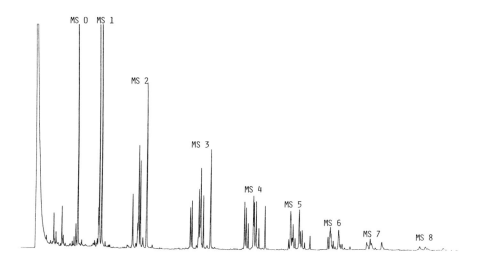

Figure 1 *Gas chromatogram of a mixture of perethyl hydroxyethyl glucitols obtained from HEC MS 2.44*

are reduced with borohydride to reduce the number of peaks in GLC.

Figure 1 shows a GLC separation of a mixture of perethyl hydroxyethyl glucitols obtained from HEC with MS 2.44. Clusters of peaks correspond to groups of monomers with the same number of ethene oxide substituents. The presence of monomers with up to 8 ethene oxide groups can be observed. The maximum number N_m of monomers for a given number of ethene oxide groups m per glucose unit amounts to $1/2(m+1)(m+2)$. Although it is possible in principle to derive the complete monomer composition from GLC, it is obvious from Figure 1 that we did not attain complete separation of monomers with MS$\geq$4. The critical factor of the method is the resolution of the GLC column. For HES with MS up to 1.2, a complete separation was obtained. Another factor is that the sample preparation for GLC is tedious and requires much skill.

3.2 Py-DCI-MS

In Py-DCI-MS, the analyte is applied to a thin metal wire and subjected to programmed rapid heating. Absorption of thermal energy causes desorption ('distillation') of the analyte from the wire. Chemical ionisation is brought about by collisions with NH_4^+ ions in a plasma which is generated by electron impact ionisation of ammonia. These ionisation conditions are very mild and result in the formation of molecular ions only. Fragments occurring in mass spectra are due exclusively to thermal degradation (pyrolysis) in the course of the heating. Being an older technique than Fast Atom Bombardment MS, Py-DCI-MS affords similar information but lacks the interferences due to matrix effects. The application of Py-DCI-MS to structural studies of carbohydrates was pioneered by Reinhold.[7]

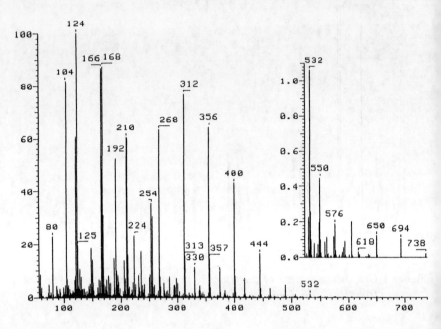

Figure 2 *Py-DCI mass spectrum of a hydrolysate of HEC MS 3.35*

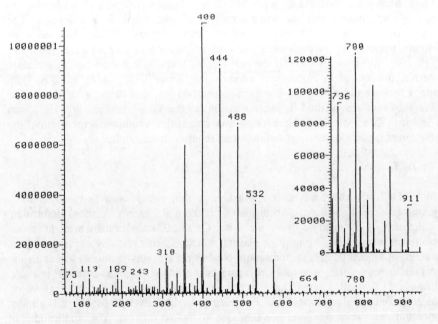

Figure 3 *Py-DCI mass spectrum of a hydrolysate of HEC MS 3.35, permethylated*

We expected Py-DCI-MS to be capable to determine the monomer mass distribution of HES and HEC.

Monomer mixtures were prepared by acid hydrolysis with 2M trifluoroacetic acid. Occasionally, hydrolysates were permethylated in order to suppress pyrolytic degradation.

Figure 2 shows the Py-DCI mass spectrum for a hydrolysate of HEC with MS 3.35. A family of signals m/z [180+44m] Da (m = 0, 1, 2, etc.) represents monomer molecular ions [G + m CH$_2$CH$_2$O - H$_2$O + NH$_4$]$^+$, where G denotes glucose and m the number of ethene oxide substituents. Monomer units containing up to 9 ethene oxide groups are observed. An interesting result was the occurrence of a second family of ions m/z [80+44m] and [78+44m] Da. These signals represent chains of poly(ethene glycol) and poly(ethene glycolaldehyde) respectively, which both originate from the fission of complete poly(ethene oxide) chains from the glucose unit. The mechanism of pyrolytic degradation of HEC has been elucidated by Arisz and Boon.[8] This particular sample of HEC is seen to contain poly(ethene oxide) chains with up to 5 ethene oxide units.

The temperature of desorption of the sample from the wire can be lowered by permethylation. In this way, fragmentation by pyrolysis was prevented. The mass spectrum of Figure 3 features only molecular ions.

It can be concluded that Py-DCI-MS provides information on monomer mass distribution of HES and HEC as well as on the length distribution of poly(ethene oxide) chains with minimal sample preparation. The latter information is difficult to obtain otherwise, because of the insufficient resolution of GLC at higher MS. Due to the occurrence of thermal degradation, quantitative results are not expected. Our results show that Py-DCI-MS is a powerful technique for rapid fingerprinting of HES and HEC and that this method yields information which is complementary to that obtained from GLC.

4 CLUSTERING TENDENCY OF SUBSTITUENTS IN PARTIALLY METHYLATED STARCH

Starch granules are semi-crystalline. The crystalline lamellae are arranged tangentially with a spacing of c. 10 nm. These lamellae are built up by the radially oriented short side chains of amylopectin, which are organised in double helices.[9] In this way, a single amylopectin molecule traverses a large number of alternate crystalline and amorphous domains.[10]

Recently, we presented evidence that methylation of granular starch occurs preferentially in the amorphous regions.[2] This would imply a tendency for substituents to cluster in these domains. To prove this we attempted an indirect method: sequential degradation with α-amylase and amyloglucosidase, followed by preparative separation and structural analysis of liberated oligosaccharides. Its use is based on the assumption that only glucosidic bonds between two unsubstituted glucose units are susceptible to attack by these enzymes. A starch derivative with a clustering of substituents would then yield a larger amount of glucose than randomly substituted starch. Moreover, the non-degradable high-molecular mass fraction would have a higher MS in case of a blockwise substitution pattern. This follows from the argument that a starch derivative with clustered substituents

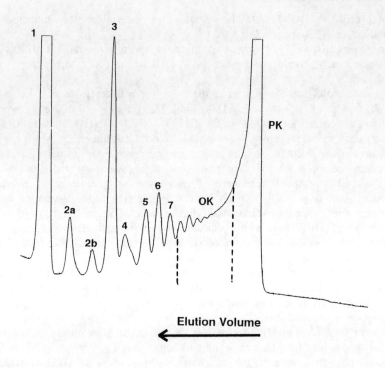

Figure 4 *GPC elution profile on Biogel P2 of enzymically degraded granular methyl starch MS 0.31*

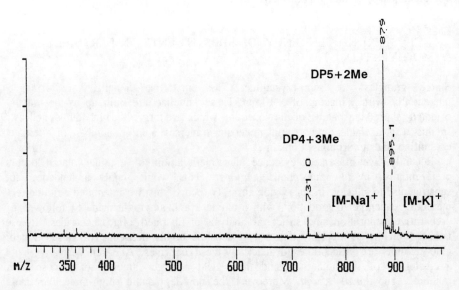

Figure 5 *MALDI TOF mass spectrum of Fraction 6 of enzymically degraded methyl starch MS 0.31 prepared in solution*

contains a larger number of glucosidic bonds between unsubstituted glucose units.

Both predictions were tested by comparing the composition of oligosaccharides obtained by enzymic degradation of starch methylated in solution (random substitution) and in granular suspension (blockwise substitution). Oligosaccharide fractions were separated by preparative GPC on Biogel P2, and the MS of these fractions was analysed by GLC. Indeed, a tendency for clustering of susbstituents was observed for starch methylated in suspension.[3]

GLC analysis only yields the average composition of oligosaccharide fractions, but it gives neither information on the number of components present in each fraction, nor on the precise location of the methyl groups. In this contribution, we take a closer look at the composition of these fractions by means of MALDI-TOF-MS.

4.1 MALDI-TOF-MS

MALDI-TOF-MS was developed less than ten years ago[11] and is now one of the most powerful tools available for the study of molecular mass distributions in complex mixtures. The analyte is mixed with an excess of a crystalline, low-molecular mass 'matrix' and is irradiated with intermittent high-energy UV pulses of very short duration from a nitrogen laser. The energy from the laser pulses is absorbed preferentially by the matrix, which 'explodes' from the target and carries the analyte molecules along with it, which are ionised simultaneously. The function of the matrix is to protect the analyte from fragmentation and, at the same time, to dilute it sufficiently to prevent its association. The matrix has to be matched judiciously to the analyte as well as to the irradiation used. For the analysis of oligosaccharides, 2,5-DHB is the matrix of choice.[12] The ion packets are accelerated to a predetermined level of kinetic energy by an electric potential, and are then allowed to travel through a field-free space with a velocity $\sim m^{-1/2}$, where they are separated and detected by their difference in flight time. The success of the method relies on the absence of fragmentation, the large mass range (> 100 kDa), and the relatively simple design of the instrument.

Figure 4 shows the GPC elution profile of the enzymic digest of granular methyl starch with MS 0.31 together with a labelling of fractions selected for further study. The MALDI-TOF mass spectrum of Fraction 6 is presented in Figure 5. This fraction is composed of two oligosaccharides: a large amount of a maltopentaose with two methyl groups (DP5+2Me) and a small amount of a maltotetraose with three methyl groups (DP4+3Me). Both components are represented by two peaks in the mass spectrum: the $[M-Na]^+$ and the $[M-K]^+$ ions. These adducts result from traces of acetate buffer used in the enzymic degradation, which have escaped the desalting step. The occurrence of these two components in the same fraction suggests that a larger number of methyl groups promotes elution from the GPC column. A similar elution behaviour was also noted for Fractions 4 and 7 (Table 1), which confirms that this phenomenon is typical for separations on Biogel P2. No evidence for fragmentation was found; peaks at $m/z < 500$ Da are due to the matrix.

An overview of the composition of the low-molecular mass fractions 2a - 7 (Figure 5) is presented in Table 1 (Fraction 1 is unsubstituted glucose). It is noted that the majority of fractions (1, 2a, 2b, 3, and 5) contain a single component. The remaining fractions consist of at most two components. This would mean that the

Table 1 *Composition of low-molecular mass fractions of enzymically degraded methyl starch MS 0.31*

Fraction Number	Suspension Reaction				Solution Reaction			
	DP	S^a	$MS_{MS}{}^a$	$MS_{GC}{}^a$	DP	S	MS_{MS}	MS_{GC}
2a	2	0	0.00	0.01	2	0	0.00	0.01
2b	2	1	0.50	0.47	2	1	0.50	0.48
3	3	1	0.33	0.35	3	1	0.33	0.34
4	3	2			3	2		
	4	1	0.31	0.35	4	1	0.40	0.34
5	4	2	0.50	0.48	4	2	0.50	0.49
6	4	3			4	3		
	5	2	0.46	0.44	5	2	0.42	0.42
7	5	3			5	3		
	6	2	0.49	0.48	6	2	0.45	0.45

^{a}: S = number of methyl groups in oligosaccharide; MS_{MS}, MS_{GC} = MS determined by MALDI-TOF-MS and GLC respectively.

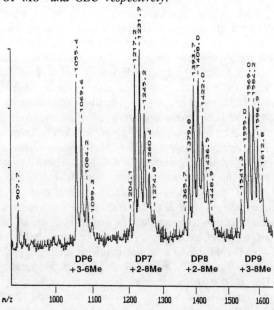

Figure 6 *Expanded scale section of the MALDI-TOF mass spectrum of the 'OK' fraction of enzymically degraded granular methyl starch MS 0.31*

Table 2 *Composition of oligosaccharides of selected DP in the 'OK' and 'PK' fractions of enzymically degraded methyl starch MS 0.31*

DP	S^a	$MS_{MS}{}^a$	$MS_{GC}{}^a$	S	MS_{MS}	MS_{GC}
		Suspension Reaction			Solution Reaction	
'OK' Fraction						
5	4	0.80	0.55	4	0.80	0.49
8	5	0.62		4	0.50	
12	7	0.58		7	0.58	
18				8	0.44	
'PK' Fraction						
12	9-10	0.79	0.57			0.46
16	11	0.69		9-10	0.59	
20	15	0.75		10-11	0.52	
24	17	0.71		13	0.54	
27	20	0.74		15	0.56	
32				17	0.53	
41				20	0.49	

^{a}: S = number of methyl groups in most abundant oligosaccharide; MS_{MS} = MS of most abundant oligosaccharide as determined by MALDI-TOF-MS; MS_{GC} = MS of 'OK' or 'PK' fraction as determined by GLC.

separation on Biogel P2 was remarkably efficient for our purpose, because sequencing of oligosaccharides should be preferably attempted with pure substances rather than with mixtures. Therefore, Fractions 2b, 3, and 5 are obvious candidates for sequencing, *e.g.* by Py-DCI-MS. Furthermore, Table 1 shows that the results of MALDI-TOF-MS are also quantitatively in agreement with those of GLC.

The real power of MALDI-TOF-MS is revealed in the analysis of complex mixtures. Figure 6 shows part of the spectrum of Fraction 'OK' at expanded scale. The mixture of a large number of different partially methylated oligosaccharides is completely resolved according to Degree of Polymerisation (DP) as well as to the number of methyl groups, whereas GLC only gives averages for MS and DP (the latter after methylation analysis).

Table 2 shows a selection of MALDI-TOF-MS data for the oligosaccharide (OK) and polysaccharide (PK) fractions of digests of starch methylated in suspension and in solution. The clustering of substituent groups in granular methyl starch is reflected in the higher number of methyl groups at a given DP. This finding agrees with the results from GLC. At present, we have no explanation for the fact that the MS of high-molecular mass oligosaccharides as measured by MALDI-TOF-MS is much higher than the corresponding MS from GLC, especially for granular methyl starch.

Another interesting feature, which was not detected by GLC, is that the non-degradable high-molecular mass fraction extends to much higher values of DP for starch methylated in solution than for granular methyl starch. Most probably, this is related to the smaller number of glucosidic bonds between unsubstituted glucose units in the former product.

The conclusion is that MALDI-TOF-MS is a powerful technique for the study of substitution patterns because of its virtually unlimited mass range, the lack of fragmentation which enables the study of complex mixtures and which, at the same time, facilitates the interpretation of mass spectra, and finally because of its modest requirements posed on sample preparation.

Acknowledgements

We gratefully acknowledge the financial support from the Netherlands Programme for Innovation Oriented Carbohydrate Research (IOP-k) and from Aqualon B.V., Zwijndrecht, The Netherlands.

References

1. O. B. Wurzburg (Ed.),'Modified Starches: Properties and Uses', CRC Press, Boca Raton, 1986.
2. P. A. M. Steeneken and E. Smith, *Carbohydr. Res.*, 1991, **209**, 239.
3. P. A. M. Steeneken and A. J. J. Woortman, *Carbohydr. Res.*, 1994, **258**, 207.
4. H. J. Lortz, *Anal. Chem.*, 1956, **28**, 892.
5. O. Larm, K. Larsson, and O. Theander, *Starch*, 1981, **33**, 240.
6. I. Ciucanu and F. Kerek, *Carbohydr. Res.*, 1984, **131**, 209.
7. V. N. Reinhold, *Methods Enzymol.*, 1987, **138**, 59.
8. P. W. Arisz and J. J. Boon, *J. Anal. Appl. Pyrolysis*, 1993, **25**, 371.
9. A. Imberty and S. Pérez, *Biopolymers*, 1988, **27**, 1205.
10. G. T. Oostergetel and E. F. J. van Bruggen, *Carbohydr. Polym.*, 1993, **21**, 7.
11. M. Karas and F. Hillenkamp, *Anal. Chem.*, 1988, **60**, 2299.
12. D. J. Harvey, *Rapid Commun. Mass Spectrom.*, 1993, **7**, 614.

STARCH: THE POLYSACCHARIDE FRACTIONS

R. F. Tester

Department of Biological Sciences (Food Science)
Glasgow Caledonian University, Southbrae Campus
Glasgow G13 1PP

1. GRANULE STRUCTURE

It is well established that starch is synthesised in the form of roughly spherical granules within cellular organelles called amyloplasts, and that these granules range in size from less than 5μm in diameter in for example rice to greater than 80μm in diameter in for example potato. In most plants there is a unimodal distribution of the granules, whilst in the *triticeae* there is a bimodal distribution comprising large A-type (circa 20μm diameter) and small B-type (circa 5μm diameter) populations. Starch granules are comprised of two major α-glucan fractions called amylose (an essentially linear molecule) and amylopectin (which is heavily branched) representing about 85-90% by weight. Certain maize mutants (sugary-1) contain a unique highly branched water soluble α-glucan called phytoglycogen.[1]

Water accounts for 10-15% of starch, with lipid which is present in cereal starches only, accounting for up to about 1%. The lipid comprises lysophospholipid (LPL) and free fatty acid (FFA) with the LPL being the almost exclusive lipid class in the *triticeae*, while other cereal starches contain high proportions of FFAs. The proportion of amylose and lipid in starches increase during granule deposition, creating an amylose and lipid concentration gradient from the hilum (centre) to the periphery of granules. The A-type granules of the *triticeae* contain proportionally more amylose than the B-type, although the smaller granules contain more lipid. Small levels of protein (<0.5% by weight) including residues of enzymes involved in starch synthesis are found even in very 'clean' (highly purified) starches. Higher levels of lipid and protein are contaminants, derived from non-starchy tissue which adsorb to the starch during extraction. More detail concerning starch composition can be found in a recent review by Morrison and Karkalas.[2]

The proportion of α-glucan that is amylopectin ranges from >95% in the low amylose or waxy starches, through 70-75% in normal starches to <30% in some high amylose starches. This amylose:amylopectin ratio is a heritable trait although there is some evidence that the ratio is affected by environment- at least in some plant species[3,4] (discussed below in more detail). Many of the physical characteristics of starches (which include gelatinisation, swelling, rheological and retrogradation properties of both solutions and gels) are directly affected by this ratio, and it is therefore important that the correct amylose dosage is chosen for a particular processing operation.

Amylose is a smaller molecule than amylopectin with a molecular weight of around 500,000, and is predominantly linear comprising α-(1-4) bonds with a very small amount of α-(1-6) bonds. The number of chains in amylose molecules from different sources range from 2-11 with the individual chain lengths containing from 250 to 670 glucose units.[2] Amylose is considered to be always amorphous in normal starch granules, although it has been reported that high amylose starches have the potential to contain amylose double helices which create crystalline structure.[5] Recent research[6,7] has indicated that in cereal starches there are in fact two amorphous forms of amylose: lipid free amylose (FAM) and lipid complexed amylose (LAM). Within LAM, the amylose is coiled around monoacyl lipids (guest molecules) located within a hydrophobic core surrounded by single polysaccharide helices (V-conformation), where the lipid accounts for about 12.5% of the complex.[6] The nature and stoichiometry of amylose-lipid complexes have been discussed in detail elsewhere[8] and will not be expanded upon here.

For many years there has been debate concerning whether inclusion complexes exist in native starch granules or are formed as a consequence of processing. Recently, proof for the existence of amylose-lipid complexes in native starch granules, has been obtained using ^{13}C-CP/MAS NMR.[6,7] This work involved comparative studies of amylose and lipid mixtures, true synthetic complexes, and native waxy, normal and high amylose starches. The fatty acid methylene carbon resonance in the rigid synthetic complexes was found to be comparable to the signals obtained from native normal and high amylose starches and was therefore a consequence of amylose-lipid complexes being formed during starch deposition. The presence of these inclusion complexes within starch granules affects their technological properties (discussed later), protects the unsaturated fatty acids from oxidation,[9] as well as making the amylose less susceptible to hydrolysis by α-amylase.[10] Interactions between true starch lipids (as opposed to surface bound lipids which are contaminants derived from non-starch sources) and other non-starch components of flour (like gluten proteins) during dough mixing are though to be impossible primarily because of amylose-lipid complexing.[11] A positive linear relationship exists between amylose and lipid in cereal starches, however, the regression equation describing the relationship varies due to species, granule size, extent of maturity and amylose dosage.[12]

The biosynthesis of starch is affected by environmental factors like growth temperature.[3,4] High temperatures depress starch deposition in barley and wheat for example, resulting in reduced yields as a consequence of fewer and small granules being synthesised. Although the amylose and lipid content of starches is mostly an heritable trait, environmental factors can also affect the proportion of these components in starches. Elevating growth temperature in barley increases the amount of lipid but tends to decrease or have no effect on the amylose content.[3] In rice starch[13] the amylose content also decreases in response to growth temperature, although in wheat, the proportion of amylose tends to increase with growth temperature.[4] The amylose content of potato starches is reported to be independent of growth temperature.[14]

It is not known why cereal starches contain lipid and the significance of either free fatty acids or lysophospholipids. The lipids might originate from the amyloplast membrane and simply be a degradation product, or play a more active and direct role in the process of starch biosynthesis and regulation of the amylose to amylopectin ratio.[15] Any proposal concerning its role in relation to the control or regulation of α-glucan synthesis in cereal starch granules must, however, take into account that in non-cereal starches lipid is essentially absent although amylose and amylopectin are co-synthesised.

Amylopectin is a very large molecule (molecular weight of many millions) which spans from the hilum (centre) to the periphery of starch granules and is composed of small linear α-(1-4) glucan chains linked together at the so called 'branch points' by α-(1-6) bonds.[16] The most peripheral (exterior) chains which contain no other chains bonded to themselves are described as A-chains. These chains are bonded to B-chains which are classified by Hizukuri[17] as B_1-B_4 depending on the number of clusters (discussed below) which they span radiating out from the hilum of the granule. According to this model, the average chain length of the A-, B_1-, B_2-, B_3- and B_4-chains are circa 12-16, 20-24, 42-48 and 69-75 respectively. There is just one C-chain per amylopectin molecule and it contains the only free reducing group. This chain is in essence the 'backbone' of the molecule with B-, and some A-chains α-(1-6) bonded to it.

The A- and small B-chains of amylopectin are found in discrete clusters within the molecule and are held together at the base of these clusters by the α-(1-6) bonds. The unit chains within the clusters readily form double helices during starch deposition which are stabilised by inter-chain hydrogen bonding. When these double helices pack together in an ordered fashion they form concentric crystalline laminates, which are interspersed with amorphous material created by α-(1-6) branching regions. It has been suggested that the forces holding starch granules together are mainly at the level of double helices and that crystallinity functions as a means of achieving dense packing rather than as a primary provider of structural stability.[18] Although the location of amylose within granules is uncertain, it has been suggested that amylose molecules are arranged radially in the granule and there is some co-crystallisation within amylopectin crystallites.[19]

The order within starch granules, as a consequence of amylopectin double helices forming and associating within crystallites, can be measured by a number of methods which essentially measure short, medium or long-range regularity. Double helices may be quantified by ^{13}C-CP/MAS-NMR.[18,20] Estimates of crystallinity within granules using this approach can overestimate the relative crystalline:amorphous ratio, since double helices need not always be associated in ordered crystalline domains. Wide angle X-ray scattering (WAXS) measures order at the level of crystallite unit cells throughout starch granules (located within crystalline laminates), and has been used to classify the helical forms (polymorphs) as A-, B- or C-type.[19,21] These helical forms (although C-type is really an intermediate form of A and B-type diffraction patterns) are characteristic of cereal, tuber and legume starches respectively. Medium range order, at the level of several repeating crystalline laminates in hydrated starch, has been measured by small angle X-ray scattering (SAXS).[22] Using this approach it has been calculated that the crystalline laminates are 6.65nm with alternating amorphous zones 2.2nm long. Longer range order as a consequence of the semi-crystalline nature of granules has traditionally been investigated using a polarising microscope. With this technique native granules appear as distorted spherocrystals with a typical dark cross. The apparent intensity of the birefringence is dependent on the thickness of the starch granule as well as on the degree of crystallinity and orientation of the crystallites.[23] Differential scanning calorimetry (DSC) measures bond breakage (primarily hydrogen bonds stabilising double helices) within starch granules when they are heated in water, and quantifies the temperature and energy (enthalpy) involved in the transition from a semi-crystalline granule to an amorphous gel. This technique can also be employed to measure higher temperature transitions in starches when amylose lipid complexes (LAM) dissociate.

1.1 Intermediate Material

It has been believed for many years that in some normal starches (especially oat and legume) there is an α-glucan fraction intermediate in molecular weight between amylose and amylopectin called 'intermediate material'.[5] Recently it has been proposed that the intermediate material identified in some starches is probably derived from hydrolysis or fragmentation of amylopectin, an artefact of the chromatographic system used to separate and quantify amylose and amylopectin or identified in amylose fractions of 'selectively precipitated α-glucans extracted from starch due to contamination with either native or hydrolysed amylopectin.[24] Although Paton[25] and others have used gel permeation chromatography (GPC) to provide proof for the existence of this intermediate fraction in solubilised oat starch, using this same technique but with a modern enzymatic and continuous post-column derivatisation system for detection,[24] it is clear that no separate intermediate material fraction exists which is distinct from amylose or amylopectin. Precipitation procedures employing 1-butanol to separate starch fractions as described by Wang and White[26] might separate α-glucans on the basis of size and branching pattern but cannot be accepted as evidence for a particular and distinct molecular species. Any polydisperse fraction like amylose or amylopectin should be able to be sub-fractionated using this precipitation procedure, however, these sub-fractions would not be intermediate in size between amylose and amylopectin if mixed and eluted together from currently available GPC systems.

1.2 Growth Rings

Growth rings, have been identified in many starches using microscopy and comprise relatively large concentric amorphous shells interspersed with relatively dense shells containing laminates of alternating crystalline and amorphous material.[23] Large starch granules such as potato and canna show pronounced growth rings when they are fully hydrated. However, as the granules dry these rings become less obvious. With smaller granules, the growth rings are much more difficult to identify and characterise. When starch granules develop in wheat plants subject to normal diurnal variation, there is one growth ring per day, although these may be absent if the plants are grown under constant illumination.[23,27] Unlike wheat, however, potato starch granules contain growth rings even if the tubers are grown under constant environmental conditions. According to French[23] growth rings represent periodic growth and with the cereal starches, daily fluctuations in carbohydrate available for starch deposition. During the day, carbohydrate products of photosynthesis are high and a dense packing of starch molecules results, while during the night the supply becomes low and packing looser.[27] Growth rings have been studied by a number of techniques including Light Microscopy (LM), Scanning Electron Microscopy (SEM) and Small Angle X-ray Scattering (SAXS) and are readily visible by SEM post chemical or enzymatic treatment. Using these techniques, it has been estimated that the dense shells are between about 120 to 400nm long containing about sixteen crystalline laminates which correspond to sixteen radiating clusters of amylopectin.[22] The relationship between the structural elements of growth rings and the distribution of amylose or lipid within starch granules is not, however, understood.

1.3 Gelatinisation and Swelling

When starch gelatinises and swells, which is primarily a property of the amylopectin fraction,[27,28] double helices in crystalline regions (and those double helices not located within ordered crystallites) uncoil and dissociate as hydrogen bonds are broken. The sugar hydroxyl groups are then free to hydrogen bond with water. Although this process has been described as 'melting', there is in fact no melting in the conventional chemical sense but uncoiling of α-glucans and expansion of the granules as water is imbibed. The relative significance of double helix versus crystallite dissociation in starch granules during gelatinisation has been discussed in detail by Cooke and Gidley.[18] They conclude that double helix dissociation is the dominant contributing factor to the enthalpy of gelatinisation rather than separation of double helices located within starch crystallites.

There are similarities between amylopectin double helix and amylose-lipid complex dissociation measured in a DSC system[8] (although for the latter, the lipid functions as a guest molecule within a single α-glucan helix), where the enthalpy change recorded is considered to be primarily due to intrahelical hydrogen bond rupture (with perhaps a small contribution from van der Waal contacts). In both these systems there will be a contribution (albeit perhaps relatively small) to the transition enthalpy by crystallite 'melting' involving breakage of some interhelical hydrogen bonds.

Each starch has its own characteristic gelatinisation onset (T_o), peak (T_p) and conclusion temperature (T_c) plus enthalpy (ΔH) of gelatinisation by DSC. This is associated with a temperature dependent swelling profile where the volume expansion (or Swelling Power) of a given starch when heated in water at different temperatures is characteristic of that starch.[28,29] Typically T_p for most starches is in the range 55-65°C, with the enthalpy of gelatinisation ranging from 10-20 J/g dry starch. The gelatinisation and swelling properties are controlled in part by amylopectin structure (unit chain lengths, intensity of branching, branching pattern, molecular weight, polydispersity and probably phosphorylation), starch composition (amylose:amylopectin ratio and lipid content which will fractionate amylose into FAM and LAM) and granular architecture (especially amylopectin crystalline:amorphous ratio), which are primarily heritable traits. However, the gelatinisation temperature (crystalline 'perfection') and enthalpy of gelatinisation (amount of hydrogen bonding within and perhaps linking amylopectin helices) can be increased by annealing,[29] or by elevating growth temperature[3,4] and this characteristic is not therefore entirely 'fixed' within the starch or completely under genetic control. It is hypothesised by this author that the effect of growth temperature on amylopectin crystallisation *in vivo* is comparable to annealing and retrogradation mechanisms which have been researched *in vitro*.

For granules to be able to expand to their maximum volume, water must be unlimiting (which is usually not the case in most food systems). Under these conditions, the onset of granular swelling is associated with the onset of gelatinisation by DSC (usually about 45°C). As the temperature is increased beyond the onset temperature, there is a progressive linear swelling region (circa 50-70°C) which eventually for most starches at higher temperatures, is followed by a plateau region. At the plateau, swollen volume does not increase markedly with temperature. Finally, however a temperature is reached where granules begin to disintegrate (at around 85°C).[28,29] When insufficient water is available to permit complete gelatinisation and unlimited swelling, the granules retain some degree of native order and are 'uncooked', as is found in for example low water systems like shortcake biscuits.[10]

When starches are heated in excess water, particularly above their respective gelatinisation temperatures, waxy starches swell more than normal starches, which in turn swell more than high amylose starches. Amylose does therefore appear in part to function as a simple diluent to amylopectin.[28,29] However, amylose also plays a more active regulatory control whereby FAM tends to decrease the gelatinisation temperature and facilitate swelling, whilst LAM acts to increase the gelatinisation temperature and restrict swelling.[6] The control and regulation of swelling exerted by the α-glucan structure and architecture within starch granules is still largely not understood.

1.4 Leaching of α-glucan

During temperature driven hydration and swelling of starch, α-glucan leaches from the granules.[28,29] The amount of leachate is usually directly proportional to the volume expansion of the granules, provided that there is no disintegration. In normal and high amylose starches this is predominantly amylose as FAM, although waxy starches also leach some α-glucan.[28] The LAM tends to remain part of the structural integrity of swollen granules and it is assumed that the amylose-lipid complex will retain its structure until the dissociation temperature of the complex is reached (circa 95-99°C for example in barley starch[6]). In hot water extracts of oat starches amylose and branched material co-leach[24] and are hence quite different from many other normal starches. It is the opinion of this author that the branched material in oat starch leachate is derived from the amylopectin fraction (as opposed to being a distinct 'intermediate material' fraction) and that some (as yet undefined) feature of these granules facilitate the solubilisation of this low molecular weight amylopectin.

1.5 Starch Damage

One major criteria of flour quality is the damaged starch content. Milling of wheat to flour induces starch damage which, at relatively low levels, is desirable because of its relative ease of hydrolysis during fermentation. Mechanical damage (discussed below) does, however, influence water holding capacity and consequently affect food product quality. Although mineral acid or α-amylase treatment of starch resulting in even a small amount of amylopectin hydrolysis reduces the swelling power,[28-30] it is usually assumed that the polysaccharide remains structurally intact during gelatinisation and swelling in most food systems. There are, however, many thermal or physical treatments (which include pyrolysis, X-ray or χ-ray irradiation and sonication) that have been shown to degrade polysaccharides[31] and in the case of thermal treatments in particular, probably play a role in modifying the structure of starchy foods.

When dry starch is subjected to physical impact and shear, in for example a conventional flour or laboratory ball-mill, a number of fractions are formed.[20,32-34] These include: native granules which have not been struck or modified; granule remnants which have suffered damage by cracking, but retain order and are optically birefringent; granule remnants which are gel forming, that is they readily hydrate and expand when placed in cold water, and; low molecular weight soluble amylopectin fragments which are generated by cleavage of amylopectin in the amorphous regions around α-(1-6) branching points. The gelatinisation temperature and enthalpy of starches are progressively reduced by impact induced damage, because of the destruction of crystallites and fragmentation of amylopectin. Temperature dependent swelling properties are also dramatically modified by damage; the effect and magnitude of which is influenced by the extent of the damage and is complicated by the fact that undamaged granules, ordered fragments and gel-forming granule fragments can swell when heated in water whilst low molecular weight material goes into solution and does not contribute to the swollen volume.[20,32-34] Amylose is not depolymerised by physical damage but leaches into solution (in the form of FAM) when native or damaged starch swell in water. The LAM is resistant to solubilisation and perhaps plays a significant role in maintaining the integrity of damaged granules.

1.6 The Future

Whilst it is true that the structure of the components of starch granules have been extensively studied and are largely understood, the way they are arranged in the granules and the effect this has on starch functionality is still very uncertain. Advances in molecular biology are continuously developing new mutants where the activities of key enzymes involved in starch biosynthesis are modified, and it should not be long before plants can be grown with starches 'tailor-made' for improving the quality of existing products and for new applications. This would reduce the need for starch manufacturers to use chemical modification post extraction to enhance or impart desirable characteristics. Other advances in the quantification of the components of starch granules and understanding of their properties, are also rapidly progressing. Developing both classical physico-chemical approaches in tandem with the genetic engineering will eventually unlock the mysteries of the nature of the organisation within starch granules.

References

1. O. E. Nelson, Genetic Control of Starch Synthesis in Maize Endosperms- A Review *in* 'New Approaches to Research on Cereal Carbohydrates', (R. D. Hill and L. Munck *eds.*), Elsevier Science Publishers, Amsterdam , 1985, p 19.

2. W. R. Morrison and J. Karkalas, Starch *in* 'Methods in Plant Biochemistry', (P. M. Dey *ed.*), Academic Press, London, 1990, Vol. 2: Starch , p 323.

3. R. F. Tester, J. B. South, W. R. Morrison and R. P. Ellis, *J. Cereal Sci.,* 1991, **13**, 113.

4. R. F. Tester, W. R. Morrison, R. H. Ellis, J. R. Piggott, G. R. Batts, T. R. Wheeler, J. I. L. Morison, P. Hadley and D. A. Ledward *J. Cereal Sci.,* 1995, **22**, 63.

5. W. Banks and C. T. Greenwood, 'Starch and its Components', Edinburgh University Press, Edinburgh, 1975, p 242.

6. W. R. Morrison, R. F. Tester, C. E. Snape, R. Law and M. J. Gidley, *Cereal Chem.,* 1993, **70**, 385.

7. W. R. Morrison, R. V. Law, and C. E. Snape, *J. Cereal Sci.*, 1993, **18**, 107.

8. J. Karkalas, S. Ma, W. R. Morrison and R. A. Pethrick, *Carbohydr. Res.*, 1995, **268**, 233.

9. L. Acker and G. Becker, *Starch/Staerk, 1971, 23, 419.*

10. J. Karkalas, R. F. Tester and W. R. Morrison, *J. Cereal Sci.*, 1992 **16**, 237.

11. L. Acker, H. J. Schmitz, and Y. Hamza, *Getreide Mehl Brot,* 1968, **18**, 45.

12. W. R. Morrison, *Cereal Foods World,* 1995, **40**, 437.

13. M. Asaoka, K. Okuno and H. Fuwa, Genetic and Environmental Control of Starch Properties in Rice Seeds *in* 'New Approaches to Research on Cereal Carbohydrates' (R. D. Hill and L. Munck *eds.*), Elsevier, Amsterdam, 1985, p29.

14. S. Hizukuri, *J. Jpn. Soc. Starch Sci.*, 1969, **17**, 73.

15. W. R. Morrison, *J. Cereal Sci.*, 1988, **8**, 1.

16. J. P. Robin, C. Mercier, R. Charbonniere and A. Guilbot, *Cereal Chem.*, 1974, **51**, 389.

17. S. Hizukuri, *Carbohydr. Res.*, 1986, **147**, 342.

18. D. Cooke and M. J. Gidley, *Carbohydr. Res.*, 1992, **227**, 103.

19. J. M. V. Blanshard, The Significance of the Structure and Function of the Starch Granule in Baked Products *in* 'Chemistry and Physics of Baking' (J. M. V Blanshard, P. J. Frazier and T Galliard *eds.*), The Royal Society of Chemistry, London, 1986, p1.

20. W. R. Morrison, R. F. Tester and M. J. Gidley, *J. Cereal Sci.*, 1994, **19**, 209.

21. J. M. V. Blanshard, Starch Granule Structure and Function: a Physicochemical Approach *in* 'Starch: Properties and Potential' (T. Galliard *ed.*), John Wiley and Sons: Chichester, 1987, p16.

22. R. E. Cameron and A. M. Donald, *Polymer,* 1992, **33**, 2628.

23. D. French, Organisation of Starch Granules *in* 'Starch Chemistry and Technology', (R. L. Whistler, J. N. BeMiller and E. F. Paschall *eds.*), Academic Press, Orlando, FL, 2nd edn., 1984, p 183.

24. R. F. Tester and J. Karkalas, *Cereal Chem.*, 1996 - accepted for publication.

25. D. Paton, *Starch/Staerke,* 1979, **31**, 184.

26. L. Z. Wang and P. J. White, *Cereal Chem.*, 1994, **3**, 263.

27. M. S. Buttrose, *J. Ultrastruc. Res.*, 1960, **4**, 231.

28. R. F. Tester and W. R. Morrison, *Cereal Chem.*, 1990, **67**, 551.

29. R. F. Tester and W. R. Morrison, *Cereal Chem.*, 1990, **67**, 558.

30. W. R. Morrison and R. F. Tester, Some Properties of the Four Major Polysaccharide Fractions in Cereal Starches *in* 'Proceedings of the 43rd Australian Cereal Chemistry Conference' (C. W. Wrigley *ed.*), Royal Australian Chemical Institute, Melbourne, 1993, p72.

31. Y. Isono, T. Kumagai and T. Watanabe, *Biosci. Biotech. Biochem.*, 1994, **58**, 1799.

32. R. F. Tester, W. R Morrison, M. J. Gidley, M. Kirkland and J. Karkalas, *J. Cereal Sci.*, 1994, **20**, 59.

33. W. R. Morrison and R. F. Tester, *J. Cereal Sci.*, 1994, **20**, 69.

34. R. F. Tester and W. R. Morrison, *J. Cereal Sci.*, 1994, **20**, 175.

INTERNAL STRUCTURE OF STARCH GRANULES REVEALED BY SCATTERING STUDIES

AM Donald, TA Waigh, PJ Jenkins, MJ Gidley*, M Debet* and A Smith†

Cavendish Laboratory,University of Cambridge, Madingley Road Cambridge CB3 OHE
*Unilever Research, Colworth House, Sharnbrook, Bedford MK44 1LQ
†John Innes Centre, Colney Lane, Norwich NR4 7UH

1 INTRODUCTION

There have been a number of studies over the years utilising various forms of scattering to unravel the internal structure of starch. The well-defined characteristic peak seen in the small angle signal (both Xray[1,2] and neutron[3]) has been related to alternating stacks of crystalline and amorphous lamellae. To model the full low angle scattering curve, these lamellae must be embedded in a second type of amorphous region which has been identified[4] with the amorphous growth ring, observed by electron microscopy[5-7].

Wide angle Xray scattering (WAXS) provides information on the local molecular packing, and as such the type of starch crystallinity, A or B. During gelatinisation of the granule, the crystallinity is lost. This can also be detected by birefringence measurements, but loss of birefringence is associated with significantly longer range ordering than the intermolecular level detected by WAXS. By monitoring the loss of crystallinity during gelatinisation, and measuring this simultaneously with the long range periodicity determined by the small angle scattering[8], it is possible to explore in considerable detail how gelatinisation is occurring. This approach is used in this paper, with application to a variety of different cultivars, and to the study of granules at different stages of maturation.

2 EXPERIMENTAL

Starches from a variety of sources have been examined. Pea mutants were supplied by Dr Cliff Hedley (John Innes Institute); waxy maize granules were extracted at different times after pollination, with samples ranging from 16-50 days after pollination; mature waxy maize was a gift from National Starch.

Small and wide angle Xray scattering (SAXS and WAXS respectively) experiments were carried out on station 8.2 at the CCLRC Daresbury Laboratory Synchrotron Radiation Source (Warrington, UK), as described previously[8]. For gelatinisation experiments, a heating rate of 5°C/minute was used, and data was binned every 12s. DSC data was collected simultaneously to permit temperature calibration. Analysis of the SAXS data was carried out using the 3 region model developed by Cameron and Donald[4], and shown in figure 1. This model contains 6 parameters, all of

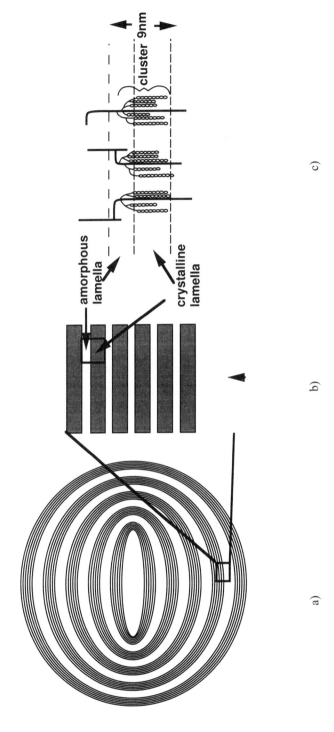

Figure 1 Model for the starch granule used to fit the SAXS data. The whole granule is shown schematically in a). Stacks of semicrystalline lamellae are separated by amorphous growth rings. b) shows a magnified view of one such stack, showing that it is made up of alternating crystalline and amorphous lamellae. c) the crystalline lamellae comprise regions of lined up double helices formed from amylopectin branches. The amorphous lamellae are where the amylopectin branch points sit.

which have a physical meaning, but during gelatinisation it is found that most of the parameters of the model remain unchanged. In particular the overall repeat distance remains constant, indicating swelling of the granule is not associated with swelling of the crystalline regions. However what do change are the two electron density differences, $\Delta\rho$ (the difference between the electron densities of the crystalline and amorphous lamellae) and $\Delta\rho_u$ (the difference between the electron densities of the amorphous lamellae and the amorphous growth rings), and the way these two vary can be fitted at all stages of the gelatinisation process, and related to the gelatinisation endotherm. These two quantities change because of the combined effects of water entering the granule and amylose leaching out, altering the local electron densities.

Simultaneous with the SAXS data, WAXS data was collected. From this it is possible to follow how the crystallinity loss proceeds during gelatinisation. The crystallinity index at each stage of gelatinisation was determined using the method of Wakelin[9]. This method allows the relative amount of crystallinity to be evaluated, by using the initial and final states as representing 100 and 0% crystallinity respectively. This does not imply that the granule starts off as 100% crystalline, but uses this as a reference state for monitoring subsequent loss of crystallinity.

Small angle neutron scattering (SANS) was carried out on the LOQ beamline at the CCLRC Rutherford Appleton Laboratory. The prime advantage of using neutrons is that they are scattered differently by hydrogen and deuterium atoms. Therefore by using water in which a proportion of D_2O has been substituted one may effectively change the scattering contrast. This provides a means of obtaining information of a type inaccessible by SAXS giving essentially absolute values for the amount of water present in the different regions of the granule during gelatinisation. Full details of the method are given in [10]. Since practically achievable fluxes are much lower for neutrons than for Xrays, it has to date only proved possible to study the gelatinisation of waxy maize starch, which has the strongest scattering of all starch types so far investigated.

3 RESULTS AND DISCUSSION

3.1 Gelatinisation of Waxy Maize

Figure 1 shows the results of gelatinising waxy maize. Figures 2a and 2b show how $\Delta\rho$ and $\Delta\rho_u$ vary with temperature, derived from fitting the SAXS data. Also shown are the onset, start, peak and conclusion temperatures of the gelatinisation endotherm, derived from DSC data. Figure 2c shows the variation in crystallinity index with temperature, derived from WAXS. From this data it is clear that crystallinity loss occurs throughout the DSC endotherm, but there is still some residual crystallinity left at the end which is lost at yet higher temperature. This finding is consistent with results of Morrison and Tester [11], who showed that swelling continued beyond the end of the endotherm. The SAXS data shows that changes first occur in the background region, since $\Delta\rho_u$ starts to fall before $\Delta\rho$. This suggests that initial swelling is primarily associated with the growth rings, with changes in the semicrystalline lamellae only occurring at higher temperatures. However, because these measurements only tell us about relative changes, SAXS on its own cannot provide a complete mixture.

These results have therefore been supplemented with SANS studies of the same system. The analysis is complex, and the method has been published elsewhere[10]. From

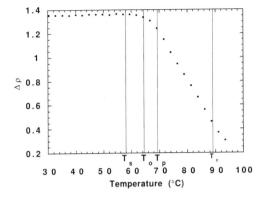

Figure 2a Variation in $\Delta\rho$ with temperature for waxy maize. Also shown are the onset, start, peak and conclusion temperatures of the DSC endotherm.

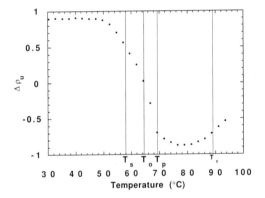

Figure 2b Variation in $\Delta\rho_u$ with temperature for waxy maize. Also shown are the onset, start, peak and conclusion temperatures of the DSC endotherm.

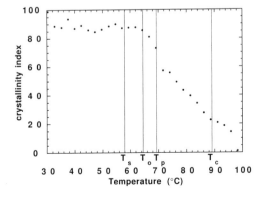

Figure 2c Variation in crystallinity index with temperature for waxy maize.

the measurements it is possible to obtain a measure of the fraction of water in the different regions of the granule. The results for the amorphous growth ring and amorphous lamellae are shown in figure 3a, and for the crystalline lamellae in figure 3b. At room temperature both types of amorphous region contain a high level of water, and much higher than in the crystalline regions. As heating proceeds, it can be seen from figure 3a that the driving force for water uptake is so high in the amorphous growth ring that it pulls water off the amorphous lamellae as swelling and gelatinisation starts to occur. The level of water in the crystalline lamellae is much lower at around 0.1,

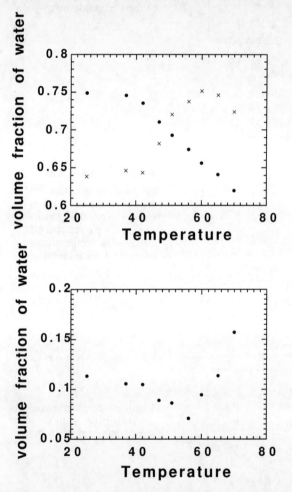

Figure 3a Variation in water volume fraction in (x) the amorphous background region and (•) the amorphous lamellae for waxy maize, derived from SANS data.

Figure 3b Variation in water volume fraction in the crystalline lamellae of waxy maize, derived from SANS data.

and remains constant until just below T_p at which point it starts to rise. Thus the SANS data confirms the SAXS data that the initial location of water uptake and swelling is in the amorphous growth ring. This finding suggests that the growth ring, which has not received much attention in the literature, may play a crucial role in determining the response of the granule to cooking.

Although only waxy maize has been studied by SANS, a variety of other species have been examined by SAX/WAXS, all showing the same trends. When interpreting the response of the granule to gelatinisation we therefore believe the results are of general validity. The model we believe best describes gelatinisation is a variant of that proposed by Donovan[12], and is primarily a swelling-driven process. Swelling starts in the amorphous growth ring, with the semicrystalline stacks essentially unchanged as shown by the constancy of the long spacing. Amylopectin molecules in the growth ring will nevertheless connect to the semicrystalline stack, and so there will be a coupling leading to a stress build-up on the lamellae as swelling proceeds. This stress will cause the amylopectin helices in the lamellae to disassociate, leading to destruction of the crystals. This picture is substantiated by results on gelatinisation in limiting water, to be presented elsewhere[13].

3.2 Waxy Maize Growth Series

In the analysis of the SAXS from starches such as described above, it is necessary to make various simplifying assumptions. One of these is that the granule is homogeneous throughout its interior. There is evidence from other work that this is not the case. Morrison, for instance, has examined the distribution of lipid through the granule and shown that there is a lower concentration at the equatorial groove[11]. It is also known that enzymatic digestion tends to be non-uniform. To examine how the granule develops as it grows, a series of samples have been examined at increasing times post-pollination. Although it is still necessary to make the assumption that at each stage the granule is homogeneous, the trends observed help to understand how the structure may vary from the inside to the outside of the granule.

Figure 4 shows a series of SAXS curves for granules of different maturity,

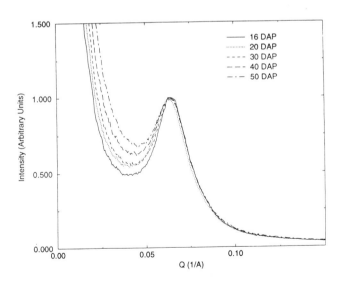

Figure 4 SAXS curves for a series of waxy maize samples, extracted at 16, 20, 30, 40 and 50 days post pollination.

each normalised so that the height of the primary peak is the same. It should be noted that within the Cameron and Donald model, this peak arises from the semicrystalline lamellar stack and the low q scattering arises from the growth rings. It can be seen that there is a systematic trend in behaviour. As the granules mature the low q scattering moves out to higher angles, equivalent to a decrease in periodicity. These results therefore show two main findings. Firstly that there is a systematic change in the internal structure as development proceeds, and the assumption that the granule is homogeneous is not a good one. Secondly, that the scattering due to the growth ring moves to higher q indicates that it is the growth rings that are changing, and that either they are getting smaller, or their density is increasing. Monte Carlo modelling predicts the latter. This finding seems to be consistent with findings that enzymatic attack often starts at the inside, since it would be expected that an enzyme will preferentially attack amorphous growth rings.

3.3 Mutant Lines of Peas

The studies of gelatinisation of wild type starches provide new insight into how processing may be affected by internal stucture of the granule. However with the growing ability to manipulate starches by genetic modification, it is equally important to know how changes altering the pathways of starch synthesis affect the granule structure. In turn this may provide new information as to what controls the way the supramolecular structure of the granule is created. To this end we have turned our attention to near isogenic lines of peas with mutations affecting enzymes on the pathway of starch synthesis. Figure 5 shows the SAXS curves for 3 mutants plus wild-type for comparison.

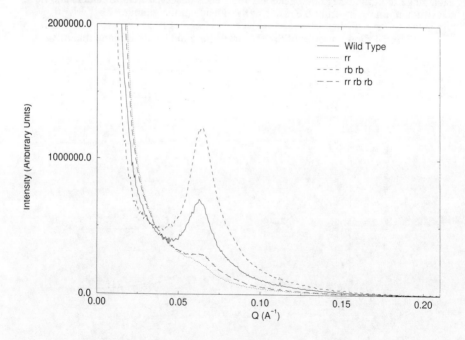

Figure 5 SAXS curves for wild-type plus 3 mutant peas.

rb rb affects ADP-glucose pyrophosphorylase and leads to a marked reduction in the overall amount of starch produced, and an increase in the amylopectin: amylose ratio. As a consequence there is a substantial increase in the overall scattering arising from the semicrystalline lamellar stack, although the basic structure of the granule seems to be the same as for wild-type. *rr* has an extreme effect on the amylopectin:amylose ratio, lowering it substantially. It reduces the activity of the starch branching enzyme by eliminating one of two isoforms. The marked reduction of the amount of amylopectin has the effect of causing the almost complete disappearance of the peak from the semicrystalline stack, since without much amylopectin there can be no crystallinity. Interestingly, the *rr rb rb* cross shows intermediate behaviour, implying that at least in this case the two effects are essentially additive.

4 CONCLUSIONS

Using this range of scattering techniques provides much quantitative information on the internal structure of starch granules, and how they break down. Clearly this has importance for understanding the processing of currently available commercial materials, but it is hoped this approach will also prove useful in contributing to the understanding of how the starch is laid down and how genetic manipulation affects this process. This step is clearly crucial if there is to be a complete logical progression from plant scientists to food processors.

ACKNOWLEDGEMENTS

The authors wish to thank the BBSRC for financial support; they acknowledge access to both the Daresbury SRS source and the Rutherford Appleton Laboratory and are grateful to the beamline scientists at both stations for their assistance. They also acknowledge Dr Cliff Hedley for the original supply of some of the mutant pea lines.

REFERENCES

1. C.J. Sterling, *J.Poly. Sci.*, 1962, **56**, S10-12.
2. G.T. Oostergetel and E.F.G.V. Bruggen, *Starke*, 1989, **41**, 331-335.
3. J.M.V. Blanshard, D.R. Bates, A.H. Muhr, D.L. Worcester, and J.S. Higgins, *Carbohydrate Polymers*, 1984, **4**, 427-442.
4. R.E. Cameron and A.M. Donald, *Polymer*, 1992, **33**, 2628-35.
5. M.S. Buttrose, *Journal of Ultrastructure Research*, 1960, **4**, 231-257.
6. M. Yamaguchi, K. Kainuma, and D. French, *J. Ultrastructure Res.*, 1979, **69**, 249-261.
7. D. French, *Organisation of starch granules.*, in *Starch: Chemstry and Technology,* R.L. Whistler, J.N. BeMiller, andE.F. Paschall, Editor. 1984, Academic Press, Inc.: London. p. 183-247.
8. P.J. Jenkins, R.E. Cameron, A.M. Donald, W. Bras, G.E. Derbyshire, G.R. Mant, and A.J. Ryan, *J Poly Sci Phys Ed.*, 1994, **32**, 1579-83.
9. J.H. Wakelin, H.S. Virgil, and E. Crystal, *J Appl Phys*, 1959, **30**, 1654-1662.
10. P.J. Jenkins and A.M. Donald, *Polymer*, 1996, **37**, 5559-68.
11. R.F. Tester and W.R. Morrison, *Cereal Chemistry*, 1990, **67**, 551.
12. J. Donovan, *Biopolymers*, 1979, **18**, 263.
13. P.J. Jenkins and A.M. Donald, *in preparation*, 1997 .

PLANT BIOTECHNOLOGY: TECHNICAL BARRIERS TO STARCH IMPROVEMENT

Peter L. Keeling, ExSeed Genetics L.L.C., 1573 Food Science Building, Iowa State University, Ames, IA 50011-1061

Telephone: (515) 294-3259   FAX: (515) 294-2644

Key words: starch synthase - maize endosperm -cDNA clones - *Zea mays*

Abstract

Biotechnological developments in starch synthesis hold great promise for modifying the basic structural composition of starch. Opportunities for structural modifications can be broken-down into a few main classes, (i) amylose and amylopectin ratios and molecular weight, (ii) amylopectin branching patterns and branched chain lengths, (iii) phosphate, lipid and protein content, and (iv) cross-linking between chains. These structural changes may be achieved by an empirical approach in which individual enzymes in the pathway are under and over expressed. However, there is evidence that this kind of approach has its limits and even with the conventional mutants we can see that there are reasons to doubt that we adequately understand the mechanism of starch assembly. In this paper I will focus on what I believe are the major technical challenges facing this new science today. In particular we are faced with a poor understanding of how the enzymes of starch biosynthesis control and contribute to starch structure. Furthermore we are also faced with trying to develop a better understanding of how starch structure and functionality are interrelated. Ultimately, it is only starch functionality that matters to the consumer and end-user/processor. Thus, we have two major technical challenges, first to improve our understanding of the links between functionality and starch fine structure, and second to establish a linkage between starch fine structure and enzymology and genetics.

1. Introduction

Agriculture uses many crop plants for food and animal feed production and for a wide variety of industrial products. Traditionally, the improvement of crop plant species involves the introduction of desired traits by genetic crosses. These breeding techniques are an extremely important component of modern agricultural practices. Recent developments in biotechnology have brought great opportunities to introduce even greater diversity into plants. Since starch is one of the main components harvested from many crops, the genetic manipulation of starch structure offers an exciting research and development opportunity for biotechnologists, food scientists, starch chemists and geneticists. Seed producers will sell improved seed to the farmer and the harvested crop will be eventually processed by end-users: a process which creates significant added-value as the product reaches the consumer. This value is driven by the consumer's interest in the functionality provided by the processor. It is important to recognize that starch not only provides calories, but also provides vital functionality to the processed material. Functionality is the key component of the added-value chain and it brings with it a significant technical challenge to researchers trying to improve the functional value of starch. This technical challenge is the prime focus of this presentation: the technical challenge of developing valuable new genetic starches.

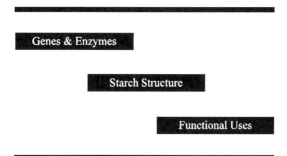

1 Starch biotechnologists face the technical challenge of forming links between disciplines

Some new genetic starches may be developed through an empirical approach involving step-by-step changes in starch genetics. However, in my opinion significant changes in value will only occur when we have improved our understanding of the links between genes & structure and structure & functionality (Figure 1). It is widely accepted that starch functionality is a consequence of the complex fine structure of amylose and amylopectin within the starch granule. Further, since this complex multidimensional structure is being formed by the enzymes responsible for starch biosynthesis, genetic manipulation is an attractive opportunity. Understanding this linkage between genetics and enzymes as well as between starch structure and functionality provides the cornerstone of the scientific efforts needed to alter starch structure. Despite the apparent obviousness of the relationship between genetics, structure and function we are still very ignorant of how these scientific disciplines link-together. Thus, although progress has been made, we do not know exactly how the genes and enzymes determine amylose and amylopectin fine structure. Similarly, the link between structure and functionality is poorly understood. Part of this ignorance stems from the very different technologies available to the different disciplines involved. For example, (i) biochemical laboratories are focussed on gene sequences and enzyme kinetics, (ii) food science laboratories deal with starch paste properties, shear forces and gel strength, (iii) starch chemists evaluate starch branching patterns and minimal chain lengths, and ultimately (iv) the consumer deals with the mouth feel, texture and shelf-life of a finished product. In my opinion, these differences between disciplines represent the key exciting opportunities, facing biotechnologists in their quest to bring-about valuable changes in starch functionality.

2. Plant Transformation

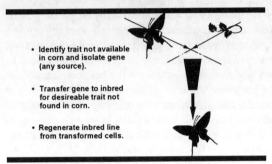

- Identify trait not available in corn and isolate gene (any source).

- Transfer gene to inbred for desireable trait not found in corn.

- Regenerate inbred line from transformed cells.

2 Plant transformation allows transfer of novel starch-synthesis genes into plants

Many different types of starches are made in various crop plants, including cereals, fruit, roots and oilseeds. Using transformation techniques it is now possible to transfer genes from one species to another. The transformation process (Figure 2) involves the key steps of first cloning useful functional genes from the donor species, next the cloned gene is repackaged in a form that permits transfer to the recipient and finally it is "injected" by various methods (for examples see Vasil, 1994) into cells from the recipient plant. Provided the new gene is incorporated into recipient plant's DNA, the introduced gene can be stably inherited and expressed. Through the transformation process we can conceive opportunities to make new specialty starches. This is a most exciting scientific endeavour, offering researchers a technical opportunity of great breadth. However, for companies intent on delivering new starches to the consumer there is no instant recipe for successfully making a valuable improvement in starch functionality.

3. Starch Biosynthesis & Molecular Biology

Starch (amylose and amylopectin) is synthesized in the plastid compartment of plant cells. In photosynthetic cells this is the chloroplast. In non-photosynthetic cells this is known as the amyloplast. We believe we know all of the enzymes responsible for starch

	Rice	Pea	Potato	Corn
ADP-glc pyrophosphorylases	**	**	**	**
Bound starch synthases	**	**	**	**
Branching enzymes	**	**	**	**
Soluble starch synthases	**	**	**	**
Amylogenin				**
Debranching enzymes				**
Disproportionating enzymes			**	

3 Enzymes involved in starch synthesis (** denotes clones available)

biosynthesis in plants. Even if this statement is correct, we do not know precisely how these enzymes and their multiple isoforms interact together, nor do we know precisely what substrate specificities the individual enzymes have and what products are produced.

At the present time the key enzymes appear to be ADP-glc pyrophosphorylases, starch synthases, branching enzymes, debranching enzymes, amylogenin and disproportionating enzymes (Figure 3). There are clones available for each of these enzymes from a variety of plant species, most notably maize, potato, pea and barley (for review see Smith and Martin, 1995; Wasserman, 1995). Furthermore there are glycogen synthesis enzymes available from bacteria and animals. Using known mutants some progress has been made in understanding the roles of some of these enzymes in starch biosynthesis. Some new starches have been made with potato transformation using antisense technology. Although similar progress with biotechnology in the major cereals has not yet been made, other opportunities have emerged. For example, by examining combinations of the known starch mutants for biochemical changes and structure-function relationships some new functionalities have been discovered (see below).

4. Opportunities for Modifications in Starch Structure

Opportunities for structural modifications can be broken-down into four categories, (i) amylose and amylopectin ratios and molecular weight, (ii) amylopectin branching patterns and branched chain lengths, (iii) phosphate, lipid and protein content, and (iv) cross-linking between chains (Figure 4). These structural changes may be achieved by

Amylose/Amylopectin content
Real amylose content
Branched chain-lengths (outer/inner)
Branched chain-length distribution
Phosphate content
Lipid/Protein association
Molecular weight

4 Opportunities for starch modification

an empirical approach in which individual enzymes in the pathway are under and over expressed. However, there is evidence that this kind of approach has its limits and even with the conventional mutants we can see that there are reasons to doubt that we adequately understand the mechanism of starch assembly. It is also clear from these studies that we should be careful not to claim that we know everything about starch fine structure. For example, although high-amylose corn has been available for many years it is only recently that we have begun to accept that the apparent increase in amylose is really driven by an increase in chain-lengths of amyopectin. Thus it is arguable that we have never seen starches with an increase in real amylose content.

5. Food Starches Today

Starch produced in amyloplasts is used to produce a wide range of food products for human and animal consumption and industrial products. Several crops are known which produce different types of starch (e.g., maize, wheat, rice, sorghum, barley, banana, apple, tomato, pear, cassava,

Corn:	Regular Yellow & White
Corn:	Waxy, Amylose
Corn:	Double Mutants
Rice:	Regular, Waxy & Amylose
Sorghum:	Regular & Waxy
Potato:	Regular & Waxy
Wheat:	Regular
Tapioca:	Regular

5 Food starches available today

potato, yam, turnip, bean or pea), each with uniquely different functionalities. The type or quality of starch makes it suitable for certain purposes, including particular methods of processing or particular end-uses. The most important food starches available today are summarized in Figure 5.

Unmodified starches have myriad of uses in everything from corrugated board, to coating and sizing paper, paperboard, adhesives, salad dressings, beer, canned foods and dry food mixes. However, these unmodified or natural starches lack some key functional properties (Figure 6) which can be partially overcome through various chemical and physical modifications. These modifications

Gelatinization properties
Viscosity characteristics
Texture characteristics
Emulsification properties
Retrogradation characteristics
Temperature characteristics
Clarity of solutions
Binding properties

6 Properties lacking in unmodified starches

can provide significant added value to the starch. However, despite the important contributions made by these modifications, many natural starches have properties that cannot easily be duplicated by chemical and physical means.

Food labeling laws permit use of the generic term "modified-starch" to cover all such starch modifications. These starches are more correctly termed hydroxyethylated, hydroxypropylated, cationic, starch acetates, starch succinates, starch phosphates, pregelatinized, bleached and dextrins and maltodextrins. Whilst such starches have been available for many years in a variety of applications

Icings/Fillings - sweet goods, cookies, mixes
Frozen & canned entrees - sauces, batters
Processed vegetables - cream sauces
Condiments & sauces, gravies, salad dressings
Flavor encapsulation
Desserts & toppings
Confectionery, powdered sugars, cake mixes

7 Food use categories of modified starches

(Figure 7), they are expensive to produce and an international trend with consumers for "natural" food products presents a great opportunity for genetically-modified starches.

The improved properties provided by the modified starches include improved emulsification, stabilization, thickening, texture, water and fat binding, oil resistance and moisture absorbance and flavor carrying. Whilst these properties do not readily translate into a recipe for designing genetic starches, we do know some general rules. For example, an increase in apparent amylose (really caused by longer chain lengths in amylopectin) appears to be related to greater gel strength, whilst more amylopectin provides greater viscosity. In addition, the amylose/amylopectin ratio is linked to starch texture and mouth feel.

6. Novel Genetic Starches

In maize there are some novel genetic starch types caused by mutations in individual steps in the pathway of starch biosynthesis. The most widely used types are known as waxy and amylose extender, with increased amylopectin and high amylose starches. Even greater diversity of functionality is known by combining different starch mutations, for example, US Patent Serial Numbers 4789557, 4790997, 4774328, 4770710, 4798735, 4767849, 4801470, 4789738, 4792458 and 5009911 describe food product uses of the maize double-mutants, producing starches of differing fine structure. Although known mutants produce altered starch, some of these lines are not suitable for crop breeding and/or for the farmers' purposes. For example, they often give relatively poor yields and can have poor germination. It is clear from these examples that improved starches may be produced by genetic manipulation of plants known to possess other favorable characteristics.

By manipulating the expression of one or more starch-synthesizing enzyme genes, it is possible to alter the amount and/or type of starch produced in a plant. One or more enzyme gene constructs, which may be of plant, fungal, bacterial or animal origin, may be incorporated into the plant genome by sexual crossing or by transformation. The enzyme gene may be an additional copy of the wild-type gene or may encode a modified or allelic or alternative enzyme with improved properties. Incorporation of the enzyme-gene construct(s) may have varying effects on starch structure depending on the amount and type of enzyme-gene(s) introduced.

7. Mutants and Biochemical Changes in Starch Synthesis

In order to try to better understand starch structure and functionality and the enzymology involved, we have studied changes in enzyme activities in great detail in the known mutants of corn. Using a gene-dosage series we have measured the enzymes involved in the pathway in order to gain a better understanding of how starch is deposited and how the entire pathway interacts together to make starch. Since endosperm is triploid it is possible to intercross each of the mutants individually and study starch pathway enzymes with 0, 1, 2

	SuS	AGP	STS	BE	DeBE
bt2	98%	8%	602%	250%	60%
ae	112%	351%	105%	17%	-
dull	107%	458%	126%	161%	-
su 1	92%	95%	26%	97%	42%
sh 1	9%	175%	189%	110%	-

8 Percent of wild-type enzyme activity in endosperm of selected mutants (SuS - sucrose synthase; AGP - ADPglc pyrophosphorylase; STS - soluble starch synthase; BE - branching enzyme; DeBE - debranching enzyme) induced by fully recessive mutants

and 3 doses of mutant allele. Thus, when the mutants are crossed with wild-type plants, the inheritance patterns of the mutant gene and wild-type gene depends on whether the gene is paternal or maternal in origin. For example in endosperm cells, where there are three doses of each allele, if the mutant is selfed the cells will all inherit 3 mutant genes (mmm). If instead the wild-type plant is used as a source of pollen for the mutant then there will be 2 doses of mutant allele and 1 dose of wild type allele (mm+). With the mutant as the pollen source, and the wild-type as the female there will be 2 doses of wild-type allele and 1 dose of mutant allele (++m). Selfing the wild type gives 3 wild-type alleles (+++). With this gene dosage series we have been surprised by the considerable degree of over expression seen for some enzymes (Figure 8). This effect on enzyme over expression appears to be linked with changes in flux caused by the mutation. For example, with the mutant bt2 we see over expression of starch synthase and branching enzyme. With ae mutants we see over expression of ADP-glc pyrophosphorylase. The only mutant that did not reveal any pleiotropic response was the mutant su1. Interestingly, this was the only case where there was no significant change in starch flux. In the cases of the intermediate dosages of enzymes we have seen no alteration in starch structure or functionality, and similarly no change in starch content either. From such studies we can conclude that some level of starch enzyme loss can be tolerated without a significant loss in starch yield. However, more modest changes in enzyme levels do not result in significant changes in starch structure either. It is tempting to conclude that changes in starch strucutre are only possible when accompanied by changes in starch content.

With the bt2 mutant we see a progressive loss in measurable ADPGlc pyrophosphorylase activity which correlates well with a loss in starch synthesis in the grain. This mutation does not appreciably alter starch structure. When the mutations are with sugary, dull, waxy and amylose extender we now detect changes in starch fine structure (branched chain length changes as well as changes in amylose/amylopectin ratios). In these cases there is more minor control of flux to starch. In all of these cases it is the changes in ratios of the starch synthases, debranching enzymes and branching enzymes which have resulted in alterations in starch fine structure. Through our studies we found that not only does the mutation reduce expression of key enzymes, but also it induces an overexpression of other enzymes in the pathway. Furthermore, it is only in the full mutant genotypes where we see changes in starch fine structure. Thus we can conclude that the changes in starch structure occur only when there is a significant loss in one enzyme isoform in combination with over expression of another enzyme isoform. Whilst not wishing to be bound by this proposal, these data illustrate the possibility that starch structure may be influenced not only by reducing expression of an enzyme (e.g. using antisense constructs) but also by simultaneously increasing expression of an other (e.g. using sense constructs).

8. Intermutants and Screens for Functionality

Using mutations available in maize we have used gene-dosage to effect a change in ratios of activities of isoforms of specific enzymes. This work relies on the well known fact that endosperm tissue is triploid by virtue of receiving two haploid nuclei from the maternal side and one haploid nucleus from the paternal side. The mutations called shrunken2 (sh2), brittle2 (bt2), dull (du), sugary (su), waxy (wx) and amylose extender (ae) encode isoforms of ADP glucose

Single Mutants (waxy, amylose, sugary)
- limited functionality

Double Mutants (waxy/amylose, waxy/dull)
- new functionalities, poor yield, poor germination

Gene-dose single mutants (aeaeAe or aeAeAe)
- no changes in starch structure or functionality

Inter-mutants (wxwxWx/AeAeae)
- novel functionality, normal starch yields

9 Useful mutants in starch synthesis

pyrophosphorylase, soluble starch synthase, debranching enzyme, bound starch synthase and branching enzyme. We have recently shown that some new starch functionalities are produced by inter-crossing multiple combinations of different mutants. Thus, with each mutation the pollen source may be homozygous recessive for one mutant allele whilst the female may be homozygous recessive for another mutant allele. For example, with the pollen source as waxy mutant (wxwx) and female amylose extender (aeae), the resulting endosperm tissue will be aeae+/wx++. We have termed this new combination-mutant an "*Inter-Mutant*" in order to distinguish this type from single and double mutants.

Starches from these Inter-mutants were screened for changes in starch functionality. A wide range of tests were used on all starches in order to make-up a matrix of different functionalities. The tests used included, differential scanning calorimetry, Brookfields, Brabender viscosity, starch granule size distribution, extracted starch yield and grain phenotype. Time and space does not permit a very detailed presentation

	Male				
	su2	ae	su1	wx	du
Female					
su2	****	***	*	*****	*
ae	****	**	***	*****	*
su1	*	**	**	***	**
wx	****	*****	**	**	**
du	**	**	*	*	*

10 Summary of changes in starch functionality measured in Inter-mutant combinations (number of * denotes significance of novel functionality)

of all of these data. Thus a simplified overview is presented in Figure 10. Overall, we found that some Inter-mutant combinations were more effective than others in generating a valuable new functionality.

9. Concluding remarks

Throughout this report I have emphasized the technical challenges of realizing valuable improvements in starch functionality through biotechnology. Based on studies of known mutants and starches from diverse botanical origin, there is evidence that changes in starch structure and functionality can be made. In my view these challenges present a boundless opportunity for biotechnologists, food scientists, starch chemists and geneticists. The years ahead look promising and exciting.

10. Acknowledgements

We thank Dr Ming Chang, Ed Wilhelm and Darcy Breyfogle for their outstanding support in maize breeding and field work. Thanks also to Dr George Singletary and Roshie Banisadr for excellent laboratory work. We also thank Dr Fran Katz, Dr Bob Freidman and Dr Richard Hauber for collaborating with starch functionality tests . This research was supported by Zeneca Seeds and ExSeed Genetics and American Maize Products.

11. References

Martin C.J. and Smith A.M. (1995) Starch Biosynthesis. The Plant Cell **7**: 971-985

Vasil I.K. (1994) Molecular improvement of cereals. Plant Molecular Biology **25**: 925-937.

Wasserman B.P., Harn C., Mu-Forster C. and Huang R. (1995) Progress toward genetically modified starches . Cereal Foods World **40**: 810-817

MANIPULATION OF STARCH QUALITY IN PEAS

T. L. Wang, L. Barber, J. Craig, K. Denyer, C. Harrison, J. R. Lloyd, M. MacLeod, A. Smith, C. L. Hedley

Department of Applied Genetics
John Innes Centre
Norwich NR4 7UH
UK

1 INTRODUCTION

Peas are not recognised as a starch crop. As legumes, their seeds are a rich source of protein which represents approximately one quarter of the dry weight of the seed. The uses of pea, which rely on this protein to supplement the carbohydrate from cereals, are mainly as an animal feedstuff and for human nutrition (see Figure 1). Nevertheless, starch is by far the greatest component of the seed (twice the amount of protein), but it has been largely overlooked as a useful commodity. Lowering the starch content of the seed, as occurs in wrinkled seeds (wild type peas have round seeds), has already proved useful. This led indirectly to a change in the use of the pea for human consumption, when wrinkled-seeded peas were developed for freezing, thereby opening up a significant new market. Such peas not only taste sweeter due to a decreased starch content and increased sugar content, but possess a starch that is very different to the original cultivars. This difference has alerted researchers and industry to the possibility that legumes, especially peas, could be used to provide starch as well as protein for agrofood industries. Failure to make use of this starch is to underuse the pea and to undervalue it.

The pea has been utilised for centuries as an experimental organism. By modern standards, what it lacks in suitability for use in routine gene isolation and transformation, is readily made up for in its ease of use for genetics, biochemistry and physiology; it is inbreeding, has a large seed, can be crossed easily and is relatively free of substances that interfere with biochemical reactions non-specifically. Furthermore, it is readily susceptible to chemical mutagenesis and hence it is easy to obtain mutants, providing one has the appropriate means of selection. It is upon such mutants and their effects on starch that this article will concentrate.

2 THE FROZEN PEA - MENDEL'S r LOCUS

The r mutation was thought to have occurred spontaneously around 1600[1] and is believed to have been introduced into pea varieties at the end of the eighteenth century by Knight, 1799.[2] The mutation causes the seed to be wrinkled (hence r from the latin rugosus meaning wrinkled), but there are a number of pleiotropic effects,[3] most

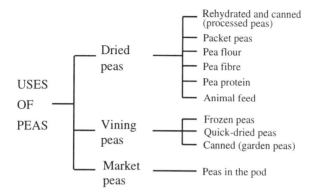

Figure 1 *The uses of peas*

notably the decrease in starch content and the increase in sweetness. The latter characteristic is the reason why varieties incorporating the *r* mutation are preferred for the frozen pea market.

The decrease in starch is associated with a change in the composition of the starch due to the primary effect of the mutation. The gene encoding starch branching enzyme I (or A)[4] is disrupted by a transposon-like insertion and hence no activity is present in mutant embryos.[5,6] The lack of enzyme activity decreases the amount of amylopectin present in the starch since these enzymes are responsible for producing branched glucans. Starch grain shape is also affected; the grains contain fissures in the mutant and are described as compound or complex, as opposed to simple or smooth in the rounded-seeded pea.

3 A SECOND *RUGOSUS* MUTANT, *rb*

In the early 1960s, Kooistra[7] examined the wrinkled-seeded variety Cennia, a recent Dutch introduction, and discovered it had smooth starch grains. He therefore went on to examine 900 wrinkled-seeded pea varieties, some of which had similar characteristics to Cennia, and concluded that they had all descended from an American introduction Alsweet. Genetic analysis indicated that the mutation in Cennia and Alsweet was due to a second *rugosus* locus which he named *rb*. The *rb* mutation decreases the activity of the enzyme ADPG pyrophosphorylase by ca. 90%.[4,8] This enzyme supplies the starch synthases with the precursor, ADPglucose. In the mutant there is a decrease in the starch content of the seed to the same extent as in the *r* mutant, but an increase in the amylopectin content of the starch in comparison with the wild type (*RRRbRb*). As will be mentioned later, it appears that decreasing the amount of precursor available to the synthases decreases amylose content. The starch grains in seeds of the mutant are smaller, as is the case for *rr* lines, but they are simple like the wild type and not like *rr*.

Kooistra also examined the phenotype of the double mutant (*rrrbrb*) and showed that the starch was decreased further, but the grains were like the *rr* line. Using near-isolines of the *r* and *rb* mutants, we have confirmed this and shown that (a) the starch grains are smaller and (b) that the amylose content is approximately midway between that of the other two mutants[9] (Table 1).

3 NEW *RUGOSUS* MUTANTS

Until recently, therefore, there existed only two *rugosus* loci and only one form of each mutant gene (allele) at each locus.   Detailed investigations of the physiology, biochemistry, cell and molecular biology associated with different aspects of the phenotype of near-isolines of the *r* mutant, however, led to a model linking all the characteristics of the mutant.[3]   A similar model could be used to explain the characteristics of *rb* lines. A prediction from this model was that the seed of all mutants in which there was a decrease in starch content would be wrinkled. The reverse would not necessarily be true, since the driving force for the wrinkling is the additional uptake, and subsequent loss during drying, of water by the embryo.   This uptake is a consequence of an increase in osmotic pressure in the cells due to their higher than normal sugar content.   Any similar imbalance in osmotic relations in the seed, even in the absence of a change in starch, would lead to a wrinkled seed.

Based on this model, therefore, we initiated a mutation programme to obtain additional wrinkled-seeded mutants using chemical mutagenesis of imbibed seed. This programme was extremely successful and led to the isolation of more than 30 lines bearing wrinkled seeds.[10]   Crosses between the majority of these lines showed that 5 complementation groups existed. (When lines within a complementation group are crossed, the seed produced will be wrinkled, that from crosses between groups will be round.)   Within each group, it has been established that there are a number of alleles[11] (Table 2).   This is important to know since the recovery of multiple alleles indicates that we have saturated the screen and have obtained mutants representing all loci influencing the wrinkling of the seed. It is also important since null mutants are likely to exist in each group which helps to determine true nature of the phenotype. Furthermore, it helps in the analysis of the structure of the protein encoded at the mutant locus, indicating regions important for conserving enzyme activity (see below). The new loci have been termed *rug-3*, *rug-4* and *rug-5* to indicate the morphology of the seed, to ensure correct nomenclatures and to prevent confusion between alleles[11].

Table 1 *Storage product composition of the seeds of* r *and* rb *near-isolines*

Genotype	Starch (% dry wt.)	Amylose (% of starch)	Protein (% dry wt.)	Lipid (% dry wt.)
RRRbRb	54.2	32.7	22.1	1.9
rrRbRb	36.1	65.1	25.8	3.6
RRrbrb	35.5	23.0	26.7	4.5
rrrbrb	23.5	49.0	28.0	5.1

Table 2 *Gene symbols for pea* rugosus *mutants by complementation group*

1	2	3	4	5
r	rb	$rug\text{-}3^a$	$rug\text{-}4^a$	$rug\text{-}5^a$
r^c	rb^c	$rug\text{-}3^b$	$rug\text{-}4^b$	$rug\text{-}5^b$
r^d	rb^d	$rug\text{-}3^c$	$rug\text{-}4^c$	$rug\text{-}5^c$
r^e	rb^e	$rug\text{-}3^d$		
r^f	rb^f	$rug\text{-}3^e$		
r^g	rb^g			
r^h	rb^h			
r^i	rb^i			
r^j				
r^k				

From our recent investigations, perhaps the most useful feature to the breeder or 'user' of these mutants is the additional variation in starch content and composition, even for existing loci. Table 3 shows the range of starch and amylose contents for the 5 loci where this point is illustrated clearly by mutants at the r locus. A detailed examination of these mutants[12] has shown that one (r^f) is a null mutant, producing no starch branching enzyme I. Seed of this mutant has the lowest starch content. Two other mutants show no significant difference from this line, but within the whole group, three other groups exist with starch contents significantly different from each other. The weakest effect of any of the mutations (r^g) gives a starch content of almost 37%. A molecular analysis of these lines also indicates that the latter mutation was generated in the least conserved region of the protein analysed to date.[12]

The five mutants at the *rug-3* locus show a phenotype in which the seed contains very small amounts of starch. The one line ($rug\text{-}3^a$) in which a significant percentage (12%) of starch is present has also been used to show that amylose is present in these lines, though initially it had been thought that this was not the case.[10] The phenotype is also manifest elsewhere in the plant since iodine staining of alcohol-cleared leaves gives no blue coloration for any of the lines. Similar mutants have also been found in *Arabidopsis*[13,14] and tobacco,[15] but these in pea are the first reported in a crop plant. The *rug-4* and *rug-5* mutations both lead to a decrease in the starch content of the seed, *rug-5* more so than *rug-4*. An analysis of enzyme activities in *rug-4* mutant seeds has shown that sucrose synthase activity is decreased in all these mutants.[16] If this is indeed the case, then other products and processes dependent on sucrose are likely to be affected. This is supported by the fact that nodule function appears to be curtailed in *rug-4* plants[16] and extremely pale green weak plants are produced in the absence of nitrogen fertiliser. In *rug-5* mutants, the activity of a 77kDa starch synthase is decreased. This synthase is considered to contribute most of the

Table 3 *The range of starch contents in* rugosus *pea mutants*

Lines	Starch content (% of dry weight)	Amylose content (% of starch)
Wild type (BC1/RR)	50-55	30-35
r	27-36	60-75
rb	30-37	23-32
rug-3	1-12	12
rug-4	38-43	31-33
rug-5	29-36	43-52

activity in the wild-type pea seed.[17]   Although the amylose content of *rug-4* mutants falls within the range of the wild type, the *rug-5* mutants show a significant increase over the wild type. The starch grain shape is also severely altered (Figure 2). Mutants of this phenotype have not been reported before.

From the known effect of the *rb* mutation and that predicted for *rug-4*, mutants at these loci interfere with the supply of precursors to the synthases. In these mutants there appears to be a linear relationship between the amount of starch and its amylose content. Since data for *rug-3* mutants also fit this correlation, one would predict a similar function for the *Rug-3* protein in supplying precursors to the synthases.

4 ROUND-SEEDED MUTANTS: the *lam* locus

In the original screening for pea seed mutants, a wrinkled-seeded character was used. It is likely that this screen was saturated for the reasons mentioned above. It is highly unlikely, however, that the screen would permit the isolation of all mutants in the starch pathway since *rugosus* mutants only represent those mutants with a lowered amount of starch in the dry seed. In order to select mutants in the starch biosynthetic pathway

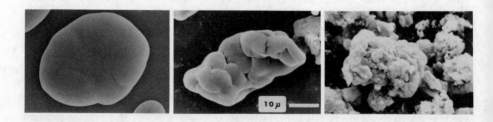

Figure 2 *Starch grains from wild type (left),* rug-5 *(centre) and* rb/rug-5 *double mutant pea seeds*

where there is no decrease in starch, for example pea equivalents to the maize *waxy* mutants,[18] a different screen is required. To this end, an additional screen[19] was carried out on mutagenised material based on the staining properties of the starch in the presence of iodine as used to isolate the *amf* mutant of potato. In this screen, the seed coat and cotyledon surface of round-seeded peas was removed and starch grains transferred to filters which were subsequently stained with iodine solution. The screen led to isolation of 5 mutants at a new locus in peas, named *lam* (*low am*ylose).[19] All these mutants possessed starch which stained brown rather than blue on the filters. An analysis of the starch in these seeds indicated little or no change in content, confirming the original hypothesis. The amylose content of the starch was very low or none depending on the way in which the measurement was made. At this point a slight digression is worthwhile. We routinely make measurements of starch and amylose content using a DMSO extraction method[10] and the data in Tables 1 and 3 are values obtained using this method. If, however, a perchloric acid extraction method[20,21] is used for the amylose, lower values are invariably obtained. Using this method, *lam* mutants show no amylose in their starch. Gel filtration chromatography also indicated that these mutants have no amylose.[19] A biochemical analysis of *lam* mutants showed that they lacked granule-bound starch synthase I activity and in several mutants the enzyme was undetectable.[19] The isolation of mutants at this locus means that 6 loci have now been identified in peas that affect the starch composition of the seed. These loci are listed in Table 4 with their equivalents in maize and potato.

5 MUTANT COMBINATIONS

The original study by Kooistra investigated not only the effects of the newly-discovered *rb* locus, but also the effects of combining *r* and *rb* mutations in the same line (see above). The discovery of 3 new loci means that further combinations can be made, both between *rugosus* mutants and between *rugosus* and *lam* mutants. A preliminary analysis of double *rugosus*[9] and double *rugosus/lam*[22] mutants has indicated that further changes to starch content and composition can be generated. This is clear from some of the starch grains produced by these mutants, Figure 2 showing a starch grain from a *rb/rug-5* mutant. Combinations of *rugosus* mutants always resulted in amounts of

Table 4 *Starch mutants*

Pea	Maize	Potato	Enzyme activity affected
r	*ae*	-	starch branching enzyme A
rb	*sh2*	-	ADPG pyrophosphorylase
rug-3	-	-	unknown
rug-4	*sh1,sus*	-	sucrose synthase
rug-5	-	-	starch synthase (GB)SSII
lam	*wx*	*amf*	starch synthase (GB)SSI

starch in the seed lower than either parent and those involving *r* mutants always gave amylose contents greater than the wild type.

Perhaps the most interesting combinations, however, are the *rugosus/lam* combinations. Since GBSSI is considered to be responsible for amylose synthesis one would anticipate that combinations involving *lam* null mutants would result in a lack of amylose in the starch i.e. that the *lam* phenotype overrides all others. This is stated to be the case for the maize *wx* mutants[18]. In pea, however, a substantial percentage of amylose is present in combinations with *rug-5* and *r* as measured following DMSO extraction or by gel filtration chromatography. Whether this material is truly amylose or a similar 'intermediate' material to that noted in *wx* double mutants[18] remains to be determined, although its iodine staining properties are very similar to amylose.

6 CONCLUSIONS

The pea is an extremely useful genetic tool as illustrated in the previous sections. Six different loci have now been identified that influence the starch content of the embryo. By manipulating the seed genetically, we have been able to enhance greatly the variation for the amount of starch and its composition. Furthermore, in the process, a great deal of information has been gained about starch synthesis in the seed. It remains to be seen, however, whether the changes induced will have practical consequences. To determine these consequences we need to understand the functional properties of the new starches and, to that end, the physical properties of the starch from these mutants is being examined currently. The reader is referred to the paper by Bogracheva *et al* in this volume for details.

References

1. H. Lamprecht, Agri Hort. Genet., 1956, **14**, 1.
2. T. A. Knight, reported in O. E. White, Proc. Am. Philos. Soc., 1917, **56**, 487.
3. T. L. Wang and C. L. Hedley, Seed Sci. Res., 1991, **1**, 3.
4. C. Martin and A. M. Smith, Plant Cell, 1995, **7**, 971.
5. A. M. Smith, Planta, 1988, **175**, 270.
6. M. K. Bhattacharyya, A. M. Smith, T. H. N. Ellis, C. Hedley and C. Martin, Cell, 1990, **60**, 115.
7. E. Kooistra, Euphytica, 1962, **11**, 357.
8. A. M. Smith, M. Bettey and I. D. Bedford, Plant Physiol., 1989, **89**, 1279.
9. J. R. Lloyd, PhD Thesis, University of East Anglia, 1995.
10. T. L. Wang, A. Hadavizideh, A. Harwood, T. J. Welham, W. A. Harwood, R. Faulks and C. L. Hedley, Plant Breed., 1990, **105**, 311.
11. T. L. Wang and C. L. Hedley, Pisum Genet., 1993, **25**, 64.
12. M. MacLeod, PhD Thesis, University of East Anglia, 1994.
13. T. Caspar, S. C. Huber and C. Somerville, Plant Physiol., 1985, **79**, 11.
14. T. P. Lin, T. Caspar, C. R. Somerville and J. Preiss, Plant Physiol., 1988, **88**, 1175.
15. K. R. Hanson and N. A. McHale, Plant Physiol., 1988, **88**, 838.
16. J. Craig, unpublished data.
17. A. M. Smith, Planta, 1990, **182**, 599.
18. O. Nelson and D. Pan, Annu. Rev. Plant. Physiol. Plant Mol. Biol., 1995, **46**, 475.

19. K. Denyer, L. M. Barber, R. Burton, C. L. Hedley, C. M. Hylton, S. Johnson, D. A. Jones, J. Marshall, A. M. Smith, H. Tatge, K. Tomlinson and T. L. Wang, Plant Cell Environ., 1995, **18**, 1019.
20. J. H. M. Hovenkamp-Hermelink, J. N. De Vries, P. Adamse, E. Jacobsen, B. Witholt and W. J. Feenstra, Potato Res., 1988, **31**, 241.
21. N.U. Haase and W. Kempf, Stärke, 1990, **42**, 294.
22. T. L. Wang and L. M. Barber, unpublished data.

CLONING, CHARACTERIZATION AND MODIFICATION OF GENES ENCODING STARCH BRANCHING ENZYMES IN BARLEY

Christer Jansson, Sun Chuanxin, Sathish Puthigae,
Anna Deiber and Staffan Ahlandsberg

Department of Biochemistry,
The Arrhenius Laboratories,
University of Stockholm,
Sweden

1. INTRODUCTION

1.1 Starch as an industrial raw material

Starch is an important raw material for industrial applications, both for food and non-food purposes (Table 1), and the market is increasing[1-3]. Of particular interest is the use of starch as a non-petroleum chemical feedstock for the manufacture of biodegradable polymers such as "plastics", and as a non-cellulose feedstock in the paper industry. Since starch is not only totally biodegradable but also inexpensive, renewable and readily available, the replacement of certain petroleum products with starch-based products has spurred an intense research activity towards an increased usage of starch as an industrial raw material. Annual EC starch production is around 10 Mtons and grows at 4-5% annually, with 80% isolated from cereals and 20% from potatoes. Of this, 45% is used as either native or modified starch in the food, polymer, paper, chemical, building, and pharmaceutical industries, the rest being used as starch hydrolysates. The Swedish annual (1992) consumption of starch amounts to 130 ktons for non-food production (mainly paper industry) and to 37 ktons for food purposes. During the last decades the demand for starch in the Swedish paper industry has grown with 5-10% annually[4]. The markets for starch in the polymer and paper industries have significant growth potentials, especially taking into account the increasing environmental awareness[3].

1.1.1 Barley as a starch source Barley is an important crop and starch source in Europe, particularly in the Nordic countries. The average annual production (1990-1992) is 172 MT worldwide and 79 Mtons in the EC, 63% of which is starch. With annual production of 2 Mtons barley is the most common cereal in crop in Sweden. The industry for processing of barley is at hand and very well developed. The storage and handling characteristics for barley seeds are superior to those of potato. This suggests that production costs, and thereby shelf prices, for bio-engineered barley starch will be competitive. Barley is also a sturdy plant that can grow in marginal areas. Finally, barley is a diploid and genetically flexible, and the knowledge in barley genetics and breeding is high.

1.2 Starch biochemistry

A large portion of the photosynthetically assimilated carbon in plants is channeled into the biosynthesis of the energy stores starch and sucrose, by far the two most widely used chemicals in the food industry. Starch is a biopolymer and serves as the major storage carbohydrate in plants. It is found primarily in seeds, tubers and roots, where it is deposited as granules in the amyloplast (a differentiated plastid organelle). During energy mobilization, starch is catabolized to glucose and fed into the glycolysis pathway.

Table 1. *Industrial use of Starch and Starch Derivatives.*

State	Industry	Application
Polymeric Starch (Amylose, Amylopectin)	Food	Thickeners, extenders, texturants, low-calorie snacks
	Paper	Beater sizing, surface sizing, coating
	Textile	Warp sizing, finishing, printing
	Polymer	Biodegradable "plastics", graft-polymers, absorbents, phenolics, adhesives
Hydrolysis products (Glucose, Maltose, Fructose, Dextrins)	Biotechnology	Cyclic dextrins as drug supports, glucose as feedstock for enzymes, hormones, antibiotics, vaccines
	Chemical	Glucose as feedstock for organic acids, organic solvents
	Fermentation and Brewing	Ethanol, liquors, spirits, beer, malt products
	Food and Beverage	Sweeteners and stabilizing agents

Starch is a mixture of amylose and amylopectin, both glucose polymers. Amylose is a mostly linear polymer of 200-2000 α-1,4 bonded glucose moieties with rare α-1,6 branch points. Amylopectin, on the other hand, is highly α-1,6 branched, with a complex structure of 10^6-10^8 MW and up to 3 x 10^6 glucose subunits, making it one of the largest biological molecules in nature. In the plant, starch is deposited as starch granules in chloroplasts of photosynthetic tissues or in amyloplasts of endosperm (seeds), tubers and roots. In most plants, starch consists of 20-30% amylose and 70-80% amylopectin. The structure of the amylose and amylopectin molecules, the amylose/amylopectin ratio, the degree of substitution, and the association of lipids and proteins are responsible for the functional qualities of starch.

In photosynthetic and non-photosynthetic tissues alike, the glucose moiety of ADP-glucose is incorporated in the growing starch polymer with the help of starch synthases; Granule-bound starch synthase (GBSS) and soluble starch synthase (SSS). The formation of α-1,6 linkages in amylopectin is catalyzed by starch branching enzymes (SBE)[5,6]. The final structure of amylopectin is governed by the activities of different SBEs, starch synthases and, possibly, a debranching enzyme. Synthesis of a nascent amylopectin molecule can occur either via amylose or directly from a glucose primer + ADP-glucose.

1.3 Genetically engineered starch

Concomitant with the increased interest in starch as an industrial raw material is a growing recognition of the potential for production of modified starch in transgenic plants. This approach offers the possibility to replace much of the post-harvest separations and chemical modifications, which are sometimes environmentally hazardous, expensive, and time-consuming. It also makes possible the production, *in planta*, of novel carbohydrates. Transgenic plants producing all-amylose starch in seeds or tubers might be achieved by inactivation of all *sbe* genes (encoding SBE). Such a genotype would be difficult to obtain by conventional breeding or by random mutagenesis due to the presence of several isoforms of SBE. Selective inactivation of *sbe* genes, on the other hand, would be expected to result in starches with altered branching patterns. Examples of post-harvest modifications that could be replaced by a transgenic approach are production of cross-linked starch for increased stability, production of starch with well-defined granule sizes for the brewing industry, and substitutions such as hydroxypropylation, methylation, carboxylation and phosphorylation. Production of modified starch in the plant itself is already well known and tolerated in nature, since altered starch patterns are found in a variety of mutants[5].

2 RESULTS

2.1 Identification and isolation of SBE forms

Using a variety of branching enzyme assays and FPLC (Fast Protein Liquid Chromatography) protein fractionation procedures we have identified four different forms of SBE in barley (Figure 1). We have purified one SBE form to homogeneity as a 51/50 kDa polypeptide[7] and partly purified two SBEII forms. The 51/50 kDa enzyme represents a novel SBE form[7] and, therefore, its branching enzyme activity was established by several, independent measurements such as phosphorylation stimulation assays, amylose branching assays, native gel assays and Thin Layer Chromatography.

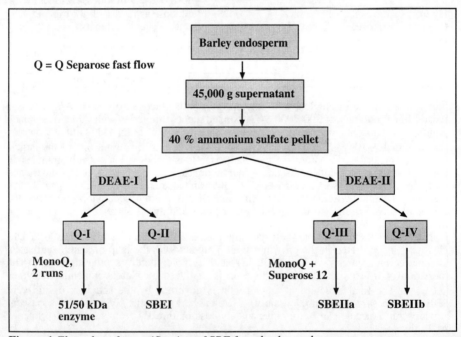

Figure 1 *Flow chart for purification of SBE from barley endosperm.*

2.1 Isolation and characterization of *sbe* genes

2.1.1 Sequence analysis Using consensus primers, RACE-PCR (Rapid Amplified cDNA Ends Polymerase Chain Reaction), RT-PCR (Reverse Transcribed PCR) and cDNA expression library screening we have isolated full-length cDNA clones of *sbe*I, *sbe*IIa and *sbe*IIb, and a partial clone of the gene for a low-molecular weight (51/50 kDa) SBE form. We have fully sequenced the cDNA clones for *sbe*IIa and *sbe*IIb and to 90% the clone for *sbe*I. We have isolated full-length genomic clones for *sbe*IIa and *sbe*IIb and sequenced the promoter region, first and second exons and the first intron for these clones. The *sbe*IIa and *sbe*IIb cDNAs are homologous but they differ in length and sequence at the 5' end. The full-length sizes are 2.7 and 2.9 nt for *sbe*IIa and *sbe*IIb,

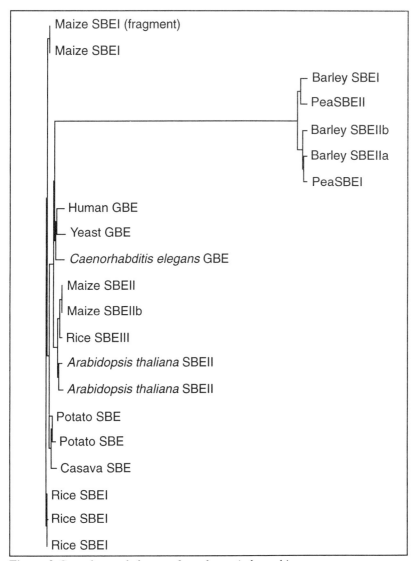

Figure 2 *Growthtree phylogram for eukaryotic branching enzymes.*

respectively. The deduced amino acid sequence for the two SBEII forms demonstrates that the N-terminal of SBEIIb is 65 amino acids longer than that of SBEIIa. Sequence identity searches showed that this N-terminal extension is homologous to regions in various membrane proteins. Thus it is conceivable that SBEIIb is bound to the amyloplast membrane (or to the starch granule). The estimated molecular mass for mature SBEI, SBEIIa and SBEIIB is 80, 80 and 85 kDa, respectively.

Alignment of the deduced amino acid sequences for barley SBEI, SBEIIa and SBEIIb and of all other branching enzyme sequences deposited in the data banks reveals that for the 34 aligned sequences there is very little similarity in the first and last 30% of the sequence whereas a significant degree of similarity can be found in the middle region.

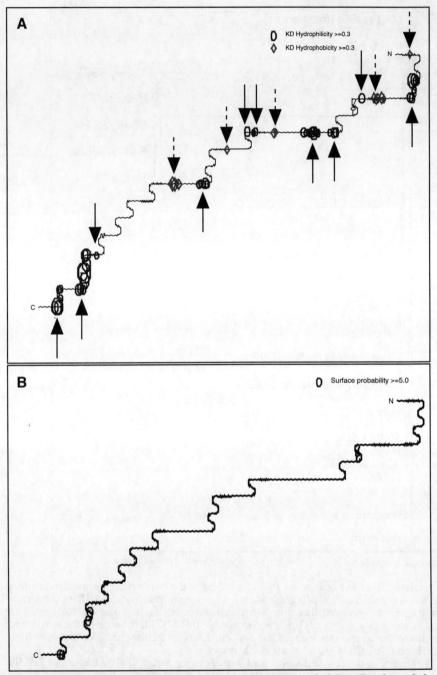

Figure 3 *Chou-Fasman hydrophobicity (A) and surface probability (B) plots of the consensus sequence for branching enzymes. A: Solid arrows indicate putative hydrophilic regions and dashed arrows putative hydrophobic regions.*

In that central portion there are several short boxes of homology, **hsHas**, **DG/AfRfDG/A**, **A/GED/EvS/Tg/a**, **A/cESHDQA/S** and **GiGF**, reading from the N- to the C-terminus. Upper case indicates 100% identity and lower case less that 100% identity (caused either by amino acid variations or introduction of gaps). X/Y denotes that a position is occupied by either X or Y, with approximately the same preference, whereas X/y indicates that the position is occupied by either X or Y with X preferred over Y. The underlined residues have been implicated as constituents of the active site in starch branching enzymes[8].

A growthtree phylogram constructed from the amino acid sequences of eukaryotic SBE forms (Figure 2) demonstrates that barley SBEI, SBEIIa and SBEIIb are closely related to each other and to pea SBEI and SBEII. These forms are set off from the other starch and glycogen branching enzymes. It should be noted that when the tree is constructed from the nucleotide sequences the outcome is somewhat different with the different barley *sbe* genes spread apart (Sathish, P. and Jansson, C., unpublished).

Chou-Fasman peptide plots of the consensus sequence for eukaryotic and prokaryotic SBE sequences (Figures 3A and B) show that the C-terminus is likely to be hydrophilic and surface-exposed.

2.1.2 Gene expression Northern blot analyses (Figure 4) suggest that expression for *sbe*IIa and *sbe*IIb peaks at around 12 days after pollination (a.p.) whereas expression for *sbe*I it peaks at around 20 days a.p. SBE activity in barley endosperm peaks at a developmental stage of 60-80 mg fresh weight per caryopsis, corresponding to approximately 16 days after flowering[7]. Based on enzyme purification data, the major portion of the SBE activity at that stage is accounted for by the 51/50 kDa form[7].

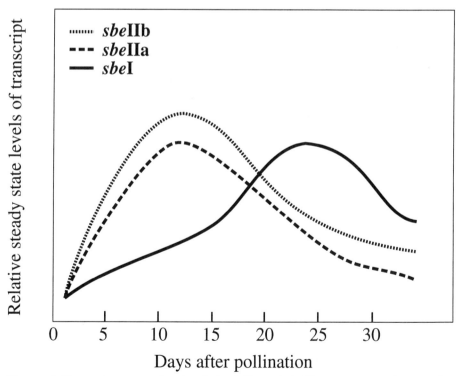

Figure 4 *Expression pattern for the sbeI, sbeIIa and sbeIIb genes assayed as steady state transcript levels.*

2.2 Transgenic studies

Immature embryos are transformed using the biolistic particle bombardment technique (PDS-1000/He, Du Pont). Expression vectors for introduction of sense and antisense constructs are being designed. In previous attempts we used the *nptII* gene (conferring kanamycin resistance) as a plant-selectable marker. However, the levels of antibiotics required (>1 mg/ml) suggested that this gene is not optimal for barley selection (Figure 5). In our present constructs (Figure 6) we use the *bar* gene (conferring bialaphos and Basta resistance) under control of the ubiquitin promoter (*ubi*)+ first intron. (The *bar* gene and the intronic *ubi* promoter were provided by Dr. Peggy Lemaux). Other components are the intronic rice actin promoter *Act1*-F (provided by Dr. Ray Wu),which is used to drive the sense or antisense construct, a multiple cloning site, the *nos* terminator and the carrier plasmid p73 conferring ampicillin resistance for selection in bacteria.

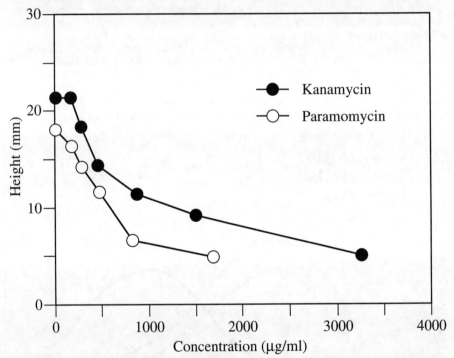

Figure 5 *Titration curves for kanamycin and paramomycin. Barley seeds were planted on media containing increasing concentrations of the antibiotics kanamycin or paramomycin sulfate. The average height of seedlings were plotted as a function of antibiotic concentration.*

2.3 Heterologous expression of branching enzyme genes in *E. coli*

Using the *gbe⁻ E. coli* strain KV832 (provided by Dr. J.A.K.W. Kiel). we have demonstrated that heterologous branching enzyme genes can be expressed in *E. coli*. The *Synechococcus* 7942 *gbe* gene (*glg B*; provided by J.A.K.W. Kiel) was cloned into the Pin-Point Xa-1 vector and used to transform *E. coli* KV40 in complementation studies. Approximately 25% of the transformants could be complemented based on the iodine staining assay.

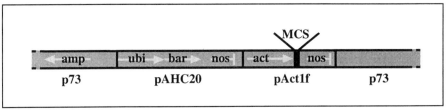

Figure 6 *Expression vector for introduction of sense and antisense constructs into barley. The orientation of the various genes is indicated by arrows.*

References

1.   Kishore, G.M. and Sommerville, C.R. *Curr. Opin. Biotechnol.*, 1993, **3**, 152-158.
2.   Shewmaker, C.K. and Stalker, D.M. *Plant Physiol.*, 1992, **100**, 1083-1086.
3.   Visser, K. and Jocobsen, E. *Trends in Biotechnol.*, 1993, **11**, 63-68.
4.   Wennerholm, T. *Aktuellt* - Sveriges lantbruksuniversitet, ALA-gruppen, 1995
5.   Sathish, P. Sun, C., Lönneborg, A. and Jansson, C. *Progress in Botany,* 1995, **56**, 301-318.
6.   Jansson, C., Sathish, P., Sun, C. and Grenthe, C. *J. Mol. Sci. - Pure & Applied Chem.,* 1995, **A32**, 895-898.
7.   Sun, C., Sathish, P., Ek, B., Deiber, A. and Jansson, C. *Physiol. Plant.*, 1996, **96**, 474-483.
8.   Martin, C. and Smith, A. M. *The Plant Cell*, 1995, **7**, 971-985.

TEMPERATURE DEPENDENCE OF SOLUBLE STARCH SYNTHASE IN THE STORAGE TISSUES OF SEEDS

C. F. Jenner and R. Sharma

Department of Plant Science
Waite Agricultural Research Institute
Glen Osmond South Australia
Australia 5064

1 INTRODUCTION

High temperature during the grain-filling stage of development diminishes yield of temperate cereals such as wheat and barley.[1] The main effect of high temperature after anthesis is on grain size, and reduced starch content accounts for the small size of the grains at high temperature.[2] High temperature impairs the conversion of sucrose to starch in the developing endosperm of wheat[3] and barley.[4] In wheat the temperature optima for the rate of grain filling[5] and for the rate of starch production in developing endosperm[3] are low, in the region of 30°C. Q_{10} values for starch deposition are also low, between 1.2 and 1.4 in the temperature range of 15°-30°C. Above 30°C, Q_{10} is smaller than one[3] because the rate of starch synthesis declines as temperature is raised further.

Differences between species in the temperature responses of dry matter (mostly starch) accumulation in seeds have been documented. Seed size in rice is reduced at high temperature less than that of wheat,[5] and the growth rate of rice seeds is more responsive to an increase of temperature in the range between 25° and 35° than is the case for wheat: seeds of rice accumulate dry matter 1.5 fold faster at 30°/25°C(day/night temperatures) compared to 24°/19°C; the corresponding ratio for wheat is 1-1.1. Variation is also evident among cultivars of wheat: Q_{10} values for the rate of grain-filling between 20° and 30°C vary from 1.01 to 1.58.[6]

Diminished starch production in wheat grains at temperatures above 30°C is associated with a reduction in extractable activity of soluble starch synthase (SSS).[7] Moreover, the synthesis of starch in the wheat endosperm at temperatures in excess of 30°C is predominantly controlled by the activity of SSS.[8,9]

The amount of SSS activity extractable from wheat endosperm does not appear to be affected by temperature in the range obetween 15° and 30°C. However, kinetic properties of the enzyme are very sensitive to temperature in this range. For example, the K_m of SSS for amylopectin increases with increase of temperature indicating that the affinity of the enzyme for one of its substrates diminishes as temperature rises.[10] The purpose of the work described in this paper is to examine the temperature response characteristics of SSS isolated from the seeds of different species, and from two cultivars of wheat known to differ in their temperature responses of starch deposition.

2 MATERIALS AND METHODS

Wheat (<u>Triticum aestivum</u> L. cvv Lyallpur and Trigo 1) was raised in a glass house and transferred to a growth room (14h 20°C day/ 10h 15°C night) 8 weeks after planting. Some plantings were transferred from the 20°/15°C room to another, identical in design, controlled at 30°C day/25°C night, nine days after anthesis. Caryopses were taken between 13 and 15 days after anthesis.

Rice (Oryza sativa cv Amaroo) was grown in a growth room (12h 30°C day/ 12h 20°C night) and caryopses were taken from panicles when the most advanced florets were 10-13 days after anthesis. Barley (Hordeum vulgare cv Schooner) was grown in a growth room (21°C day/16°C night) and florets were taken from the spike 13-15 days after anthesis.

French beans (Phaseolus vulgaris L. cv Brown Beauty) were grown in a glass house (temperature not exceeding 25°C) and the seeds were used when the cotyledons weighed *c.* 200 mg each.

2.1 Methods

About 2g fresh weight of wheat endosperms (100), barley endosperms(100), rice endosperms (200) and bean cotyledons (10) were removed from the seeds and homogenised with a pestle and mortar on ice in 10 mL of extraction buffer.[10] After centrifugation at 4°C the supernatant was brought to 40% saturation with $(NH_4)_2SO_4$ and the precipitate was desalted on a Sephadex column as described previously.[10] Soluble starch synthase was assayed as described[10] for 4 min (wheat), 5 min (bean), 2 min (rice) and 0.5 min (barley) at the temperatures indicated. For all assays the concentration of ADPglucose was 1 mM, and the concentration of amylopectin was varied from 0.05 to 40 mg mL^{-1}.

2.2 High Temperature Treatment of Wheat Grains

Wheat grains (100) were removed from the ears and placed in a beaker lined with wet filter paper. The beaker was immersed in a water bath at 35°C for 1h. An equivalent sample of grains from the same set of ears was maintained at room temperature (23°C) for 1 hr. Endosperms were isolated and extracted as described above.

3 RESULTS

Of the activity extracted from the endosperm (or cotyledons) 60-70% was recovered in the eluate from the Sephadex column. Measured activities were corrected for recovery. Throughout the range of temperature and concentrations of amylopectin investigated, the activity recorded was proportional to the volume of extract assayed (data not shown). The duration of the assays was selected such that the rate of catalysis was linear with respect to time. As the rate of catalysis during the assay of SSS from barley and rice began to decline after 1 or 2 minutes, especially at high temperature (data not shown), assay times of 0.5 and 2 minutes respectively were used for barley and rice.

Kinetic parameters were calculated by regression analyses of Eadie-Hofstee plots; typical results are shown in Figure 1 for rice. At all temperatures and for all species the data fitted straight lines. Results for crops grown at 20°C (wheat and barley) and for beans grown in the glass house are presented in Figure 2. There were substantial differences between the species, and between the two cultivars of wheat in the values of V_{max}, and the responses of V_{max} to temperature (Figure 2A). Throughout the whole temperature range V_{max} increased with increase of temperature in barley, but for wheat and bean, optima were clearly evident at temperatures of 35°-40°C and 30°C respectively.Decline in V_{max} at temperatures above the optimum is not due to high temperature irreversible inactivation of the enzyme: in the case of bean at least it has been shown that these temperature responses are reversible.[11]

In all the species surveyed (Figure 2B) K_m for amylopectin increased with increase of temperature. The greatest increases were observed in wheat (118 - 174 fold between 15°C and 45°C) and the smallest increase (11 - 15 fold) occurred in barley and bean.Compared to barley or bean, wheat had a higher affinity for amylopectin at low temperature, but a

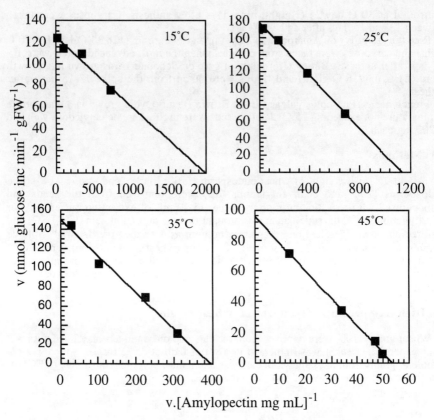

Figure 1 *Eadie - Hofstee plots for the dependence of velocity on concentration of amylopectin in rice at different temperatures*

lower affinity at high temperature.

Exposure of wheat grains, isolated from the ear, to a period (0.5h) of treatment at temperatures above 25°C reduces the amount of SSS activity extractable from the endosperm.[8] The effects of exposure of grains to 35°C for 1h on the kinetic properties of the SSS activity which survived the high temperature treatmant are compared in Figure 3 for the two cultivars of wheat. V_{max} in cv Trigo endosperm that had not been heated was higher than that in unheated cv Lyallpur. Heating reduced V_{max} of both cvs but there was no difference between them in the V_{max} of the enzyme activity that had survived heating. There were no conspicuous differences between heated and unheated grains in the affinity for amylopectin (Figure 3B).

Q_{10} for starch deposition during grain filling in wheat for the temperature interval 20°-30°C varies among cultivars.[6] Two cultivars of wheat differing in Q_{10} (20°-30°C) were included in this study, and the temperature response characteristics of SSS extracted from the endosperm of grains from plants grown at 20°C were compared to those grown at 30°C. Data for rice grown at 30°C are included for comparison with wheat grown at 30°C. V_{max} was higher for wheat grown at 20°C than that grown at 30°C (Figure 4A) throughout the entire range of temperature, but the temperature responses of V_{max} were similar at both growth temperatures. Surprisingly, the optimum temperature for V_{max} in rice (25°C) was lower than that recorded for wheat (35°-40°C). K_m for amylopectin

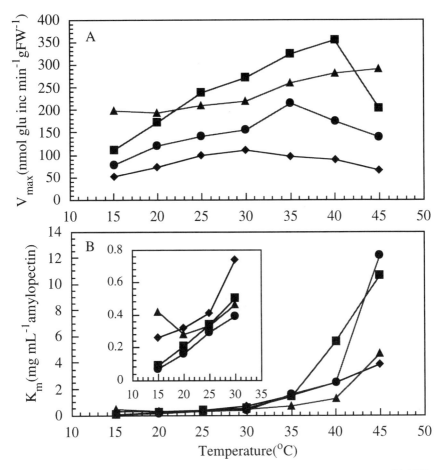

Figure 2 *Soluble Starch Synthase: Temperature responses of (A)V_{max} and (B) K_m for amylopectin in wheat cvs Trigo(■) and Lyallpur(●), barley(▲) and bean(◆)*

increased with temperature in both lots of wheat; K_m in rice changed less than in wheat with increase in temperature (Figure 4B).

Effects of temperature on reaction rate are a reflection of effects of temperature on V_{max} and K_m. V_{max} over K_m (called the efficiency of the enzyme, or the specificity constant)[12] gives further insight into the effects of temperature on the reaction catalysed by SSS. At low temperature, V_{max}/K_m was greatest for rice and least for bean (Figure 5), and with increase of temperature the values of rice and wheat fell progressively over the whole range from 15° to 45°C. Values for barley and bean appeared to display temperature optima (at 20° and 25°C respectively) and declined progressively with increase of temperature above the optima.

Wheat cv Trigo grown at 30°C had greater values of V_{max}/K_m than when grown at 20°C in the temperature range 15°-30°C. In contrast, the efficiency of SSS from Lyallpur grown at 20°C was greater than that grown at 30°C.

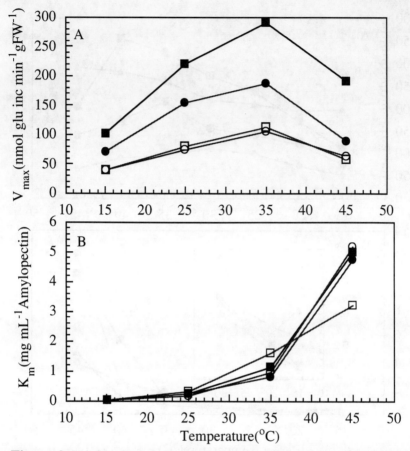

Figure 3 *Soluble Starch Synthase: Temperature responses of (A)*
V_{max} *and (B)* K_m *for amylopectin in wheat cvs Trigo(■□) and
Lyallpur(●○).The closed and open symbols represent control and
grains heated to 35°C respectively*

4 DISCUSSION

Seventy to 75% of the weight of cereal seeds is contributed by starch and protein accounts
for a further 10-15% of the dry weight. Reduction in seed weight as temperature rises
does not affect the deposition of protein. Thus reduced grain-filling at high temperature is
due to effects on the deposition of starch. The optimum temperature for the rate of starch
deposition in seeds is comparatively low (25°-35°C) and rice, adapted to sub-tropical
climates has a higher temperature optimum than wheat, a temperate crop. Variation among
cultivars in temperature tolerance of grain-filling has been documented in wheat. The main
purpose of the work described in this paper is to investigate the proposition that
temperature response characteristics of the deposition of starch in seeds is dependent upon
kinetic responses to temperature of the reaction catalysed by SSS.

Kinetic properties of SSS from all the species examined are sensitive to temperature,

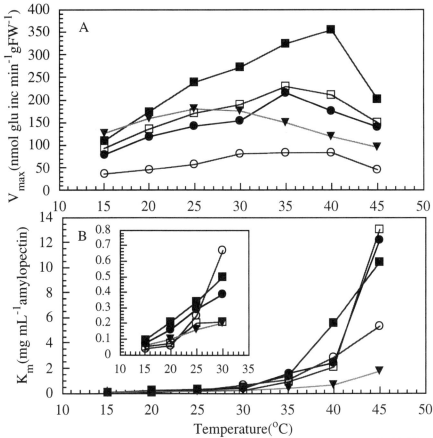

Figure 4 *Soluble starch synthase: Temperature responses of (A) V_max and (B) K_m in wheat (cvs Lyallpur ●○ and Trigo ■□) and rice(▼).The closed and open symbols in wheat represent plants grown at 20°C day/15°C night and 30°C day/25°C night respectively*

and there is considerable diversity of response to temperature among them.

4.1 Temperature and V_{max}

In only one case (barley) does V_{max} rise with temperature over the whole range (Figure 6A). For each of the other species, there is a clear optimum for maximum velocity i.e. 25°C for rice, 30°C for bean and 35° or 40°C for wheat. Although V_{max} was lower in wheat plants grown at 30°C compared to 20°C (Figure 4), the temperature response characteristics of SSS were not altered by the growth temperature (Figure 6B).

Grain-filling in rice is more tolerant of high temperature than in wheat and the Q_5 value for the rate of grain filling for the temperature interval between 25° and 30°C is estimated to be approximately 1.5 in rice but little greater than 1.0 in wheat[5].Values of $Q_5(25°-30°C)$ for V_{max} of SSS in rice and wheat (grown at 30°C) calculated from Figure 6B are

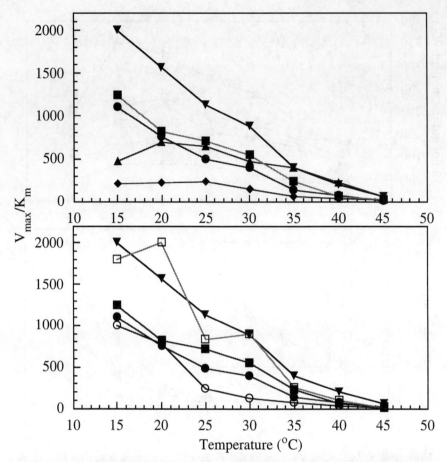

Figure 5 *Soluble starch synthase: Temperature responses of V_{max}/K_m in wheat cvs Trigo(■ □) and Lyallpur(● O), rice(▼), barley(▲) and bean(◆).The closed and open symbols in wheat represent growth temperatures of 20° day/15° night and 30° day/25° night respectively*

respectively 0.98 and 1.4 (Lyallpur) and 1.1 (Trigo). Clearly, differences between wheat and rice in their temperature responses for grain-filling are not associated with temperature responses for V_{max} of SSS.Comparison between two cultivars of wheat also shows that there is no direct link between temperature responses for grain-filling and for V_{max}. Q_{10} ($20^°$-$30^°$C) values for grain filling are 0.94 and 1.26 respectively for Lyallpur and Trigo (M. Zahedi pers. comm.) while from Figure 6 corresponding values for V_{max} are 1.29 and 1.58 for wheat grown at $20^°$C and 1.78 and 1.41 for the crop grown at $30^°$C (Table 1).

4.2 Temperature and K_m

Decline in the affinity of an enzyme for its substrate(s) with increase of temperature above the optimum for plant growth is not an uncommon response.[13] For wheat and barley (Figures 2 and 4) the highest affinities of SSS for amylopectin are observed in the

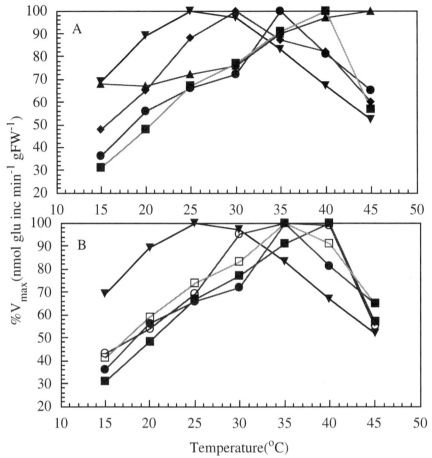

Figure 6 *Temperature responses of V*$_{max}$*for wheat cvs Trigo(■□) and Lyallpur(●O), barley(▲),bean(◆) and rice(▼).The closed and open symbols in wheat represent growth temperatures of 20°C day/15°C night and 30°C day/20°C night respectively.*

temperature range (15°-20°C) where temperate crops yield more starch. K_m for rice follows a similar trend although it does not rise as much as wheat at temperatures above 25°C.Thus there is no clear distinction between wheat (temperate) and rice (subtropical) in the temperature at which the highest affinities of SSS for amylopectin are observed.

4.3 Temperature and V_{max}/K_m

Given that V_{max}/K_m is a criterion of the efficiency of the enzyme (Figure 5), the efficiency of the rice SSS enzyme is higher than that of wheat grown at 20°C and higher than barley and bean. However, the efficiency of rice SSS falls as fast or faster with increase of temperature as it does in the other species. The temperature response of the efficiency of SSS thus does not explain the superior performance at high temperature of grain filling in rice compared to wheat. It may be worth noting however, that the efficiency of SSS from Trigo appears superior to that of Lyallpur in the temperature range

Table 1 *Temperature response characteristics of SSS from the endosperm of wheat cvs Trigo and Lyallpur grown at 20°/15°C day/night or 30°/25°C*

Assay temperature	Attribute	20°/15°C		30°/25°C	
		Trigo	Lyallpur	Trigo	Lyallpur
20°C	V_{max}	172	120	135	45
	K_m	0.21	0.16	0.07	0.06
	V_{max}/K_m	820	760	2000	780
30°C	V_{max}	272	155	190	80
	K_m	0.5	0.39	0.21	0.67
	V_{max}/K_m	542	400	900	120
20°/30°C	$Q_{10} V_{max}$	1.58	1.29	1.41	1.78

20°-30°C where Trigo performs better than Lyallpur, and that only in cv Trigo is the efficiency higher in grain grown at 30°C than it is in grain grown at 20°C.

4.4 High temperature inactivation of SSS in wheat

There are multiple isoforms of SSS in wheat endosperm,[14] separable by FPLC. Heating isolated grains for 1 hour at 35°C reduces the activity of the form(s) eluting at the higher KCl concentration relatively more than the low salt form(s)(Jenner, unpublished). Heating grains for 1h at 35°C reduced the combined activity (of all forms) by approximately 70% in Trigo and 50% in Lyallpur. However, the temperature responses of the form(s) of SSS that survived heating appeared little different to the responses of all forms combined (Figure 3). It would not appear therefore that the temperature characteristics differ among the isoforms. Nor is there any direct association between grain filling performance at high temperature (Trigo superior to Lyallpur) and survival of catalytic activity of SSS in vitro at high temperature.

4.5 Low temperature optimum for starch deposition in seeds

To what extent could kinetic characteristics of SSS account for the temperature responses of starch deposition in seeds? In general, increase of temperature results in a decrease in the efficiency of SSS for all the species examined. Moreover, in wheat at least[7] (there are no data available for other species) temperatures in excess of 30°C reduce the level of SSS activity in the endosperm, and control of the synthesis of starch is almost wholly attributable to the activity of SSS.[8,9]

Thus high temperature could act through one or both (kinetic responses and/or level of activity) of these mechanisms, to reduce the accumulation of starch. However, kinetic responses to temperature in the range between 20° and 30°C do not account for the differences between wheat and rice. Several other suppositions might be advanced as alternatives. For example, starch synthesis in rice endosperm might not be under the control of SSS as it is in wheat endosperm.[8,9]

The only kinetic temperature response feature of SSS that appears to be in accordance with Trigo's superior performance when grown at 30°C is its higher efficiency in comparison with Lyallpur (Figure 5). This is due to differences between these two cultivars in the combined effects of growth temperature on V_{max} and K_m (Table 1). One conceivable explanation for these differences is that growth at 30°C promotes the production in Trigo of a form or form(s) of SSS with high affinity for amylopectin in the temperature range between 20° and 30°C (see Figure 4).

5 SUMMARY AND CONCLUSIONS

Diversity exists between species in the temperature response characteristics of SSS catalysing the production of starch in their seeds.

Evolutionary adaptation to a warm environment (as has occurred in rice) has not been associated with the appearance of forms of SSS having kinetic responses conferring high catalytic activity at higher temperature.

Variation in temperature responses of starch synthesis among cultivars of wheat on the other hand does appear to be reflected in changes in the temperature response characteristics of SSS produced at elevated temperature.

6 ACKNOWLEDGEMENTS

This work was supported by a grant from the Australian Research Council.

References

1. I. F. Wardlaw, I. A. Dawson, P. Munibi and R. Fewster, *Aust.J. Agri. Res.,* 1989, **40**,1.
2. S. S. Bhullar and C. F. Jenner, *Aust. J. Plant Physiol.,* 1985,**12,** 363.
3. S. S. Bhullar and C. F. Jenner,_Aust. J. Plant Physiol.,* 1986, **13**, 605.
4. L. C. Macleod and C. F. Duffus, *Aust. J.Plant Physiol.,* 1988, **15**, 367.
5. T. Tashiro and I. F. Wardlaw, *Ann. Bot.,* 1989, **64**, 59.
6. L. A. Hunt, G. van der Poorten and S. Pararajasingham, *Can. J. Plant Sci.,* 1991, **71**, 609.
7. J. S. Hawker and C. F. Jenner, *Aust. J. Plant Physiol.,* 1993, **20**, 197.
8. C. F. Jenner, K. Siwek and J. S. Hawker, *Aust. J. Plant Physiol.,* 1993, **20**, 329.
9. P. L. Keeling, P. J. Bacon and D. C. Holt, *Planta,* 1993, **191**, 342.
10. C. F. Jenner, K. Denyer and J. Guerin, *Aust. J. Plant Physiol.,* 1995, **22**, 703.
11. R. Sharma and C. F. Jenner, *Proc. 45th Aust. Cereal Chem. Conf.,* 1995, 153.
12. A. Fersht, 'Enzyme Structure and Mechanism', Freeman, New York, 1985.
13. J. J. Burke, J. R. Mahan and J. L. Hatfield, *Agron. J.,* 1988, **80**, 553.
14. K. Denyer, C. M. Hylton, C. F. Jenner and A. M. Smith, *Planta,* 1995, **196**, 256.

OCCURRENCE AND EXPRESSION OF GRANULE-BOUND STARCH SYNTHASE MUTANTS IN HARD WINTER WHEAT

R. A. Graybosch, C. J. Peterson and L. E. Hansen

USDA-ARS, University of Nebraska - Lincoln
Lincoln, NE, USA, 68583

A. Hill and J. Skerritt

CSIRO, Division of Plant Industry
Canberra, ACT, Australia, 2601

1 INTRODUCTION

Granule-bound starch synthase (GBSS, EC 24.1.21) is responsible for the synthesis of amylose in the endosperm starch of cereals;[1] a role for GBSS in amylopectin synthesis also has been described for some plant species.[2] Mutations that eliminate the production of GBSS result in the amylose-free (waxy) starch. In diploid grasses such as barley (*Hordeum vulgare*) and maize (*Zea mays*) waxy mutations are common, and well characterized.[3] In common wheat (*Triticum aestivum*), an allohexaploid, there are three loci (*wxA-1, wxB-1, wxD-1*) encoding GBSS. A spontaneous waxy mutant in common wheat would require simultaneous recessive mutations at all three loci, an extremely unlikely event. To date, no spontaneously occurring waxy wheat has been isolated; waxy wheats have been produced via hybridizations of lines carrying null (non-functional) alleles at the *wx* loci.[4]

Recent advances in the purification and electrophoretic separation of wheat GBSS[5] now allow the recognition of the gene products of the three *wx* loci. In addition, Nakamura et al.[5-7] identified a number of wheat lines that carry null alleles at one of the three loci. Null alleles at *wxA-1* and *wxB-1* were found to be fairly common among wheats of Japanese and Australian origin, respectively. Only one line, Bai Huo, from China, was identified as possessing a null allele at the *wx-D1* locus. Lines carrying one or two null alleles were designated "partial-waxy" lines.[5-7] Hybridization of lines carrying the three null alleles has resulted in the production of amylose-free (waxy) common and durum wheats.[4]

Wheat breeders in the U.S. are interested in reducing or eliminating wheat starch amylose. Reduced-amylose (partial-waxy) wheats have been shown to confer superior performance in Oriental noodle applications,[8] and may confer both extended shelf-life and enhanced water absorption to baked goods. Amylose-free starch might find applications in both paper and food industries, and most likely would be useful in all present applications of waxy maize starch. Rapid development of both reduced amylose and amylose-free wheat cultivars would be facilitated by the identification of adapted lines carrying one or more *wx* null alleles. The goals of the present study were to: 1) survey recently released U.S. cultivars and advanced breeding lines to identify lines carrying *wx* null alleles, 2) to study the effects of *wx* null alleles on amylose content of hard winter wheats, and on concentrations of GBSS and 3) to compare the relative effects of environment, genotype, and genotype X environment interaction on starch amylose and GBSS contents. Our eventual goals are to develop a series of reduced-amylose and amylose-free winter wheats.

2 MATERIALS AND METHODS

The wheat lines sampled primarily were advanced experimental lines entered in 1996 USDA-ARS regional performance nurseries. A limited number of breeding lines and cultivars also were obtained directly from public and private breeding programs. The sample included 196 hard winter wheats of Great Plains origin, and 39 soft winter wheats from the eastern U.S. Grain samples were ground in a Udy cyclone mill, starch was purified, and GBSS was extracted and separated as per Nakamura et al.[5] with the following modification. GBSS protein was reduced and alkylated with 4-vinyl-pyridine, and silver stained, as per Graybosch and Morris.[9] The effects of null alleles on amylose contents were assessed using two sets of experimental materials. Experiment 1 consisted of single samples of 13 experimental lines and cultivars grown in 1994 at Berthoud, Colorado. Genotypes in Experiment 1 included wild-type (no null alleles present), *wxA-1* and *wxB-1* single nulls, and one double-null line (both *wxA-1* and *wxB-1* null alleles present). Experiment 2 consisted of 11 wheat lines grown in two replication trials at 3 Nebraska locations (Lincoln, Sidney and Scottsbluff) in both 1990 and 1991; wild type and *wxA-1* and *wxB-1* single null lines were included. Amylose contents were determined by I_2 binding after dissolution of starch in dimethylsulfoxide.[10]

The amount of GBSS protein per mg starch was determined through enzyme-linked-immunosorbent assay (ELISA). Two monoclonal antibodies (mabs) specific for GBSS[11] were used to develop a sandwich assay. Monoclonal 91484 was used as the capture antibody, and mab 91563, labeled with horse-radish peroxidase, was used as the tag (reporter). For ELISA, GBSS was extracted by boiling 10 mg starch for 10 minutes in H_2O, followed by centrifugation at 14,000 x g for 30 minutes. The resultant supernatant was diluted 1/10 before application to coated microtitre plates.

Analysis of variance and calculation of least significant differences were used to compare mean responses from both amylose determinations and ELISA. Experiment 1 was a randomized complete block design with three laboratory replications. Experiment 2 was a randomized complete block design; environment, replication within environment, and environment x line were considered random effects; line was a fixed effect in the model. Amylose contents were expressed as % amylose. A log10 transformation of ELISA optical densities (OD) was used to insure equal variances.[12]

3 RESULTS

3.1 Frequency of *wx* Null Alleles among U.S. Wheats

Null alleles at the *wx* loci were detected among U.S. hard wheats (Figure 1), and were found in materials arising from several different breeding programs (Table 1). Of 196 hard winter wheats assayed, 21 (10.7%) carried a null allele at one of the *wx* loci; 7 carried a *wx-A1* null, while 13 lines possessed a null allele at the *wx-B1* locus. One line, 'Ike' was found to carry null alleles at both *wx-A1* and *wx-B1*. The remaining hard wheats were "wild-type", that is, possessing three functional *wx* loci. Among the 39 soft wheats, only one line carrying a null allele was detected. An experimental breeding line designated LA8676-B-21-4-1-B carried a null allele at the *wx-A1* locus. No *wx-B1* nulls were found among the soft wheats, and no *wx-D1* nulls were detected in either class of wheat.

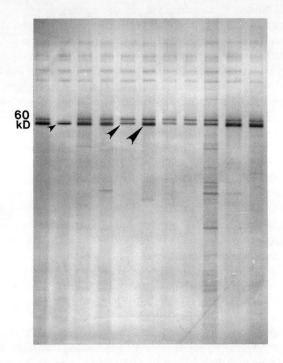

Figure 1 *Electrophoretic separation of starch proteins from several hard wheat lines; the granule-bound starch synthases (GBSS) have a M$_r$ of 60 kD. Small arrow indicates GBSS from a line (Ike) carrying null alleles at both* wxA-1 *and* wxB-1; *only the* wxD-1 *gene product is present. Medium arrow designates GBSS from TAM202* (wxB-1 *null); in TAM202 gene products of both* wxA-1 *(upper band) and* wxD-1 *are present. Large arrow indicates GBSS from a wild-type line (Redland) in which all three gene products occur.*

Table 1 *U.S. hard winter wheats found to carry wx null alleles.*

Genotype	Lines	Origin[1]
wxA-1 null	Chisholm, Cimarron	Oklahoma
	KS801072	Kansas
	Colt, NE86501	Nebraska
	Laredo, Custer	private
wxB-1 null	TAM200, TAM202, TX92V3108, TX93V5919, TX93V5922, TX93V4927	Texas
	RioBlanco, WI93335, WI93339	private
	K94H115, K94H400, K94H402	Kansas
	CO910748	Colorado
wxA-1 null, *wxB-1* null	Ike	Kansas

[1]Origin: public (University-affiliated) breeding programs are denoted by respective states; "private" indicates lines developed by commercial seed companies.

3.2 Effects of *wx* Null Alleles on Starch Amylose and GBSS Contents

3.2.1. Experiment 1. Significant differences in both amylose and GBSS contents were observed among the lines entered in Experiment 1 (Table 2). Both mean amylose and GBSS contents of Ike, the cultivar carrying null alleles at both *wxA-1* and *wxB-1,* were lower than all other lines. With the exception of Laredo, mean amylose contents of all single null lines were lower than that of Hawk and Jagger, the wild-type lines with the highest amylose contents; however, no significant differences were observed between the single null lines and the remaining wild-type lines. Among single null and wild-type lines, with the exception of Ike, a continuous range in GBSS content was observed, and few significant differences were detected. Relative to all other lines, GBSS content of Ike was markedly reduced.

3.2.2. Experiment 2. A more precise estimate of the effects of single null alleles at the *wxA-1* and *wxB-1* loci was obtained by analyzing samples of wheat lines grown at multiple locations. Analysis of variance (Table 3) demonstrated significant effects of environment, line and line X environment interaction on amylose content; hence, genetic factors (i.e. allelic status at the *wx* loci) do not explain all the variation in amylose content. The magnitude of the environment mean square, relative to the mean square of line, indicates environmental sources of variation in amylose content exceeded those of genotypic sources, in this experiment. In contrast, environment and line X environment interactions did not significantly affect GBSS contents; significant differences, however, were observed among lines (Table 3). Mean amylose contents of three of the single null lines, TAM200, TAM202 and KS801072 were significantly lower than those of all wild-type lines with the exception of TAM107 (Table 4). A similar result was obtained with the ELISA, except GBSS content of KS801072 did not differ from that of most wild-type lines.

Table 2 *Amylose and GBSS contents from 13 wheat lines obtained from Berthoud, Colorado.*

Line	Genotype	Amylose content (%)	GBSS content (ELISA OD)
Hawk	wild-type	34.04	1.59
WI89163W	wild-type	30.95	1.39
WI90540	wild-type	31.08	1.52
WI91103	wild-type	29.71	1.77
WI93357	wild-type	30.28	1.67
WI93481	wild-type	32.44	1.87
Jagger	wild-type	33.72	1.76
Laredo	null *wxA-1*	29.10	1.81
Custer	null *wxA-1*	28.12	1.74
RioBlanco	null *wxB-1*	28.70	1.37
WI93335	null *wxB-1*	28.48	1.45
WI93339	null *wxB1*	26.41	1.62
Ike	null *wxA-1*, null *wx-B1*	19.74	0.92
l.s.d.		5.16	0.24

Table 3 *Mean squares from analysis of variance of amylose and GBSS contents of 11 wheat lines grown at 6 Nebraska environments.* * = significant at p = 0.05.

Source of variation	df	Mean squares	
		amylose	GBSS (ELISA)
Environment	5	116.20*	2.96
Rep (env.)	6	15.53	1.23
Line	10	45.10*	2.25*
Line X env.	50	15.31*	1.15
Error	60	8.59	0.05

Table 4 *Mean amylose and GBSS contents of 11 wheat lines grown at 6 Nebraska environments.*

Line	Genotype	Amylose content (%)	GBSS content (ELISA OD)
Centurk 78	wild-type	30.37	2.31
Redland	wild-type	30.21	2.02
Scout 66	wild-type	29.61	2.05
Siouxland	wild-type	30.30	1.92
TAM107	wild-type	27.49	1.93
Vona	wild-type	29.79	2.01
Chisholm	null *wxA-1*	28.56	1.95
Cimarron	null *wxA-1*	29.25	2.02
KS801072	null *wxA-1*	25.17	1.92
TAM200	null *wxB-1*	25.41	1.81
TAM202	null *wxB-1*	26.81	1.72
l.s.d.		3.21	0.19

4 DISCUSSION

Null alleles at either the *wxA-1* and *wx-B1* loci were found in approximately 10% of the sampled U.S. hard winter wheats. Results of Experiments 1 and 2 suggest the designation of lines carrying single null alleles as "partial waxy"[5-7] might be inappropriate. While some of the lines carrying single null alleles displayed significantly lower amylose contents than the majority of wild-type wheats, the effect was not universal, and the amylose depression was slight. When data from the two experiments were combined, wild-type wheats averaged 30.77% amylose. The mean amylose contents of lines with single nulls at either the *wxA-1* and *wx-B1* were 28.04% and 27.16%, respectively. In contrast, starch of Ike, a line carrying null alleles at both the *wxA-1* and *wx-B1* loci, was only 19.74% amylose. The designation "partial waxy" might be more appropriately applied only to double null lines. While genetic variation did exist among single null lines for amylose content, and lines with only 25% amylose were identified, the significant effects of environment and line X environment interactions suggest such lines would be unreliable as sources of low amylose wheat starch. In this regard, production of increased numbers of adapted double null lines, through intermatings of single null wheats, will be the more likely source of low amylose.

Lines with single null alleles possess 2/3 the number of possible active genes, yet starch amylose content was reduced to, at most, 83% of that of wild-type lines. Even in Ike, a line possessing only 1 active gene, starch amylose was 67% that of wild-type, when, based on the number of active alleles, one might expect this figure to be only 33%. The ELISA data suggest that active *wx* alleles are capable of dosage compensation; that is, the amount of GBSS was not reduced proportionally to the number of inactive genes. In some lines with null alleles, GBSS content was equal to that of wild-type. The presence of a single null allele did not reduce GBSS production to an extent that enzymatic activity, and concomitant amylose contents, are markedly reduced. GBSS content of Ike was still approximately ½ that of wild-type lines. This suggests compensation in double null lines occurs both via increased protein production and enzymatic activity. Additional factors that might influence the amylose content of the various null lines would be differential enzymatic activities of the various GBSS isoforms, and background genetic effects, or, that there are more important rate-limiting steps in starch synthesis.

Despite the lack of a marked decrease in amylose content observed among lines carrying single *wx* null alleles, such lines will be extremely useful in the development of locally adapted wheats carrying reduce-amylose or amylose-free starch. Clearly, double null lines are desired for reduced-amylose starch. Intermatings of lines with single nulls will give rise to F_2 populations in which 25% of the individuals carry double nulls. Both single and double null lines also will be absolutely essential in the development of amylose-free (waxy) cultivars. Nakamura et al.[4] have produced amylose-free common wheat through intermating of Kanto107, a Japanese wheat carrying both *wxA-1* and *wxB-1* null alleles, and Bai Huo a wheat of Chinese origin carrying a *wxD-1* null. While this combination serves as the source of the amylose-free phenotype, commercial application of amylose-free wheat starch will depend upon the introgression of the trait to locally adapted cultivars. The important role of wheats with single and double nulls is demonstrated by consideration of F_2 segregation ratios. If an amylose-free wheat (triple null) is crossed to a wild-type line, only 1/64 F_2 individuals will be amylose-free. If either a single null or a double null line is used in the cross in place of wild-type, the respective frequencies increase to 1/16 and 1/4. Successful development of a new wheat cultivar generally requires the identification of the rare line, among hundreds, that possesses the requisite agronomic and quality characteristics. Identification of adapted lines with single null alleles will render this task slightly less formidable.

Breeders of hard winter wheat in the U.S. traditionally have opted to improve agronomic performance and disease resistance, and either maintain or increase quality for the production of leavened bakery products. Flour protein content and protein quality have been traditional targets for the enhancement of wheat end-use quality. However, many breeding programs in the Great Plains are now diverting significant resources to expanding the development of hard white wheats, in addition to the traditional hard red wheats. One of the anticipated uses of hard white wheat will be in the production of various types of Oriental (wet noodles). Starch amylose content has been implicated as a significant factor governing flour performance in some types of noodles.[8] The near future likely will see assays of starch amylose, and GBSS allelic composition or determination of GBSS contents via ELISA, incorporated into wheat breeding programs.

References

1. C. Martin and A.M. Smith, *The Plant Cell*, 1995, **7**, 971.
2. M.L. Maddelein, N. Libessart, F. Bellanger, B. Delrue, C. D'Hulst, N. Van den Koornhuyse, T. Fontaine, J.M. Wieruszeski, A. Decq and S. Ball, *J. Biol. Chemistry*, 1994, **269**, 25150.
3. O. Nelson and D. Pan, *Ann. Rev. Plant Physiol. Plant Mol. Biol.*, 1995, **46**, 475.
4. T. Nakamura, M. Yamamori, H. Hirano, S. Hidaka and T. Nagamine, *Mol. Gen. Genet.*, 1995, **248**, 253.
5. T. Nakamura, M. Yamamori, H. Hirano and S. Hidaka, *Plant Breeding*, 1993, **111**, 99.

6. T. Nakamura, M. Yamamori, S. Hidaka and T. Hoshino, *Japan. J. Breed.*, 1992, **42**, 681.

7. M. Yamamori, T. Nakamura, T.R. Endo and T. Nagamine, *Theor. Appl. Genet.*, 1994, **89**, 179.

8. H. Miura and S. Tanii, *Euphytica*, 1993, **72**, 171.

9. R. Graybosch and R. Morris, *J. Cereal Sci.*, 1990, **11**, 201.

10. D. Knutson and M.J. Grove, *Cereal Chemistry*, 1994, **71**, 469.

11. S. Rahman, B. Kosar-Hashemi, M.S. Samuel, A. Hill, D.C. Abbott, J.H. Skerritt, J. Preiss, R. Appels and M.K. Morell, *Aust. J. Plant Phys.*, 1995, **22**, 793.

12. R.G.D. Steel and J.H. Torrie, 'Principles and Procedures of Statistics', McGraw-Hill Book Company, New York, 1980.

THE MECHANISM OF AMYLOSE SYNTHESIS

K. Denyer, A. Edwards, C. Martin and A. M. Smith

John Innes Centre
Norwich Research Park
Colney
Norfolk NR4 7UH

1 INTRODUCTION

We know from work on the amylose free mutants of various species such as the *amf* mutant of potato,[1] the various *waxy* mutants of cereals[2-5] and the *lam* mutant of pea,[6] that amylose synthesis requires a particular isoform of starch synthase, known as granule-bound starch synthase I (GBSSI). We also know that plant storage organs contain other isoforms of starch synthase which cannot make amylose.[7] To try to understand the mechanism of amylose synthesis, we have compared the properties of the different isoforms of starch synthase. The following is a description of the diversity of isoforms in different storage organs, their locations in and around the granule and our ideas and experiments on their roles in the synthesis of amylose and amylopectin.

2 THE DIVERSITY AND LOCATIONS OF ISOFORMS OF STARCH SYNTHASE

Starch synthase activity is found both tightly bound to the starch granule and in the stroma, the soluble phase of the amyloplast surrounding the granule. It has been known for a long time that the soluble and granule-bound activities are due to different isoforms[8,9] but it is now clear that individual isoforms partition to different extents between the soluble and granule-bound fractions. The degree to which an isoform is granule-bound may depend upon its affinity for the starch granule, the period in development during which it is synthesised and its stability in the soluble fraction. GBSSI is an example of an isoform of starch synthase which is almost exclusively granule-bound.

Immunogold labelling of sections through starch granules has shown that the granule-bound isoforms exist within the granule rather than bound to the granule surface.[10] We imagine that these isoforms become buried as the granule grows. A small proportion of the starch synthases are active when buried within the granule, perhaps those which are close enough to the granule surface to receive a supply of their substrate ADPglucose. Consequently, mechanical disruption of starch granules generally results in a considerable increase in their starch synthase activity.[8, 11-13] We do not know how far in to the intact granule ADPglucose can penetrate and starch synthases can continue to elongate glucans.

Proteins other than starch synthases also become buried in starch granules. Starch-branching enzymes in developing pea embryos[10] and maize endosperm[14] have been shown to partition between the soluble and granule-bound fractions and granule-bound branching enzymes have also been identified in wheat.[15,16] There are also minor proteins within the starch granules which have not yet been identified. These proteins are likely to be enzymes involved in starch synthesis because almost all of the granule-bound proteins identified to date are enzymes which use glucan chains as a substrate. Other soluble proteins, even those involved in the production of ADPglucose such as ADPglucose pyrophosphorylase, do not become granule-bound. [14]

The diversity of types of starch synthases is illustrated in Figure 1 which compares the starch granule-bound proteins from a number of different storage organs. The granule-bound proteins were extracted from purified starch by heating to 100°C in a medium containing 2% sodium dodecyl sulfate. Many of these proteins are starch synthases. All storage organs possess GBSSI, a starch synthase of about 60 kDa. The mechanism of amylose synthesis in all of these organs is therefore likely to be similar. Storage organs differ however, in the numbers and types of isoforms of starch synthase other than GBSSI which they possess. Many possess a class of starch synthase of between 75-90kDa, antigenically related to the 77-kDa starch synthase of pea.[15,17] We will refer to starch synthases of this type as SSII. There are also starch synthases larger than SSII in potato tubers,[18] endosperm of wheat,[15] barley and maize[13] and in pea leaves.[19] Immunological studies suggest that these may represent several distinct classes of isoform of starch synthase.

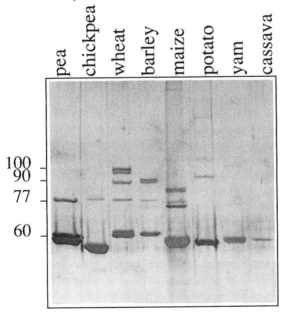

Figure 1 *Starch granule-bound proteins from cotyledons (pea, chickpea), endosperms (wheat, barley, maize) and tubers/roots (potato, yam, cassava). Molecular masses (kDa) are indicated.*

In starch from developing pea embryos there are two forms of starch synthase, GBSSI and SSII. SSII is also present in the soluble fraction and is the major soluble isoform in this storage organ, accounting for 60-70% of the soluble starch synthase activity.[20] We have investigated whether the partitioning of SSII between the soluble and granule-bound fractions is an intrinsic property of the enzyme or something peculiar to pea embryos by expressing pea-SSII in potato tubers. Comparison of soluble and insoluble (or purified starch) fractions from control and transformed plants showed that the isoform from pea partitions between the soluble phase and the granule in potato tubers in exactly the same way as it does in pea embryos.[21] The presence of SSII in the granule is therefore due to the intrinsic properties of the protein and is probably a consequence of its activity.

3 INVESTIGATION OF THE ROLES OF GBSSI AND SSII

Although GBSSI clearly makes amylose and other isoforms of starch synthase - both soluble and granule-bound - do not, the mechanism of amylose synthesis is unclear. In particular, it is not known how the products of GBSSI escape the action of starch-branching enzyme. A model was proposed to explain the differences between the roles of the isoforms based upon their different locations.[22,23] In this model, the activity of starch-branching enzyme was assumed to be confined to the soluble fraction. The product of GBSSI was said to remain unbranched because being inside the granule, it was inaccessible to active starch-branching enzyme. We now know that branching enzymes are present in the granule as well as in the soluble fraction but we do not know whether they are active in this environment. However, it is clear that this model is no longer acceptable because it does not explain why other isoforms of starch synthase within the granule cannot make amylose. This problem is particularly obvious in developing pea embryos. The *lam* mutant of pea lacks GBSSI but has granule-bound SSII. The granule-bound starch synthase activity of *lam* mutant embryos is about a third of that in wildtype embryos and yet the starch of the mutant is amylose-free as judged by gel-permeation chromatography.[6] There must be other, more subtle differences between the isoforms of starch synthase, apart from their locations, which explain the different functions.

To understand the basis for the different roles of GBSSI and SSII, we have compared the amino acid sequences of these proteins. The sequence of the pea GBSSI is similar to those of GBSSIs in other storage organs and to the C-terminal 60 kDa of SSII from pea embryos.[24] The greatest difference between GBSSI and SSII is that SSII has an extra domain of about 200 amino acids at its N-terminus. To discover the function of this domain and to compare the properties of the enzymes generally, we have expressed both the full-length proteins and truncated and chimeric proteins in *Escherichia coli*. We have shown that the N-terminal domain of SSII is not required for activity,[21] but it seems likely that it has a specific function. It may influence the binding of the enzyme to starch or its interactions with other enzymes and it may contribute to the differences between the roles of GBSSI and SSII. This work is still in progress.

The properties of isoforms of starch synthases in soluble extracts of *E. coli* may not be identical to their properties inside the starch granule. For example, GBSSI has a low specific activity in soluble extracts which may mean that the environment within the granule is required for full activity. Therefore, we complimented studies of the properties

a. Elution of amylose and amylopectin

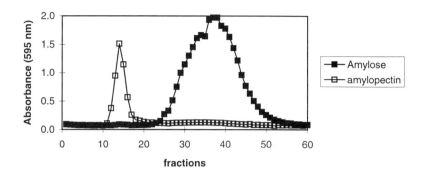

b. Elution of wildtype and mutant starches

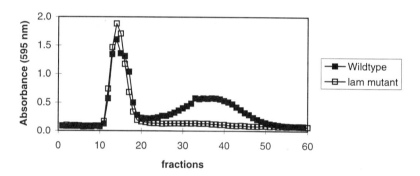

c. Elution of ^{14}C-labelled glucans

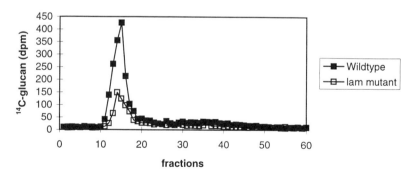

Figure 2 *The position of elution from columns of Sepharose CL2B of purified amylose and amylopectin, starch from wildtype and lam mutant pea embryos and the ^{14}C-labelled products of incubations of starch granules with ADP[U-^{14}C]glucose.*

of GBSSI and SSII in extracts of *E. coli* with a study of their properties in isolated starch granules. We wanted to know which glucan, amylose or amylopectin, was made by which isoform of starch synthase and whether the nature of the products was influenced by the incubation conditions. Granules containing active starch synthases were isolated from wildtype and *lam*-mutant pea embryos and incubated with ^{14}C-labelled ADPglucose. The alcohol-insoluble products, ^{14}C-labelled amylose and amylopectin, were separated by gel-permeation chromatography and quantified (Figure 2).

With starch from wildtype pea embryos, it might be expected that most of the ^{14}C-glucose would be incorporated into amylose since most of the granule-bound starch synthase activity is due to GBSSI, the isoform responsible for amylose synthesis. However, we found that most of the label was incorporated into amylopectin. Incorporation predominantly into amylopectin was also reported for starch granules isolated from sweet potato tubers.[25] Debranching of the labelled amylopectin from pea and sweet potato starch showed that the ^{14}C-glucose was preferentially added to the longest amylopectin chains. This may indicate that only a subset of extra-long amylopectin chains are available for elongation within the granule. Comparison of the incorporation of ^{14}C-glucose into granules from wildtype (containing GBSSI and SSII) and *lam*-mutant embryos (containing SSII only) suggested that both isoforms of starch synthase are capable of using glucose from ADPglucose to elongate amylopectin. The incorporation of [14]Cglucose into amylopectin in starch granules from the *lam* mutant was, as expected, lower than the incorporation into amylopectin in wildtype granules.

4 THE ROLE OF MALTO-OLIGOSACCHARIDES IN AMYLOSE SYNTHESIS

We reasoned that the failure of isolated granules to synthesise amylose might be because something normally present *in vivo* which was required for amylose synthesis had been washed out of the granules during their preparation. Incubating purified starch granules from pea embryos or potato tubers with soluble extracts from these organs caused a small but significant increase in the incorporation into amylose. Boiling the soluble extracts did not affect their ability to stimulate amylose synthesis suggesting that soluble enzymes were not responsible. However, treating the boiled, soluble extracts with α-glucosidase (maltase) before incubating them with the granules did destroy the stimulatory factor suggesting that it might consist of short chains of glucose units (malto-oligosaccharides). Consistent with this, incubating wildtype starch granules with 100 mM maltotriose had a dramatic effect on the pattern of incorporation of glucose from ADPglucose (Figure 3). Incorporation into amylose was stimulated whilst the incorporation into amylopectin was decreased. Maltose and malto-oligosaccharides consisting of up to seven glucose units stimulated amylose synthesis but glucose and cellobiose did not. Malto-oligosaccharides did not stimulate the elongation of amylopectin by SSII in starch granules from *lam*-mutant peas. The stimulation of amylose synthesis in starch from wildtype peas is therefore likely to be due to a specific effect of malto-oligosaccharides on GBSSI.

We are investigating the mechanism of malto-oligosaccharide-dependent stimulation of amylose synthesis. One possibility is that malto-oligosaccharides act as primers and are elongated by starch synthases. Another possibility is that malto-oligosaccharides act as effectors, stimulating starch synthases to elongate pre-existing amylose molecules. These two mechanisms are not mutually exclusive and both may operate simultaneously.

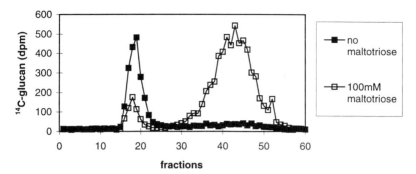

Figure 3 *The elution of ^{14}C-labelled glucans from a column of Sepharose CL2B showing the stimulation by maltotriose of amylose synthesis in starch granules from wildtype pea embryos.*

1. Synthesis of branched polymer.

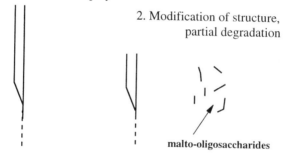

2. Modification of structure, partial degradation

malto-oligosaccharides

3. Elongation of amylopectin, synthesis of amylose

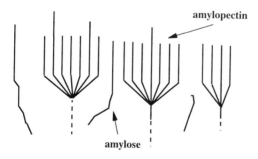

amylopectin

amylose

Figure 4 *The proposed roles of GBSSI, SSII and malto-oligosaccharides in the synthesis and elaboration of amylose and amylopectin.*

5 SUMMARY

We have shown that, in addition to GBSSI, amylose synthesis requires malto-oligosaccharide. Malto-oligosaccharides with the ability to stimulate amylose synthesis are present in extracts of plant storage organs. Although at this stage we do not know whether these factors are present inside the amyloplast, this seems likely. This discovery suggests a role for the degradative enzymes (amylases, glucan phosphorylase debranching enzymes, disproportionating enzyme, etc.) which are present in developing storage organs during starch synthesis. They may provide the malto-oligosaccharides required for amylose synthesis.

A model explaining our current understanding of the roles of GBSSI, SSII and malto-oligsaccharides is shown in Figure 4. Soluble starch synthases together with soluble starch-branching enzymes synthesise an amylopectin framework. This is modified by degradative enzymes, which in the process generate malto-oligosaccharides. Within the granule, in the presence of malto-oligosaccharides, GBSSI synthesises amylose, possibly using malto-oligosaccharides as primers. Our results and those of others[26] have shown that GBSSI is also capable of elongating amylopectin. Other granule-bound starch synthases such as SSII also elongate amylopectin. This elongation, which probably results in the production of extra-long chains may have important implications for the physical properties of starch. Extra-long chains have been implicated in determining the viscosity of pastes of amylopectin.[27]

References

1.   E. Jacobsen et. al., *Euphytica*, 1989, **44**, 43.
2.   O.E. Nelson and H.W. Rines, *Biochem. Biophys. Res. Commun.*, 1962, **9**, 297.
3.   J.S. Hseih, *Bot. Bull. Acad. Sinica*, 1988, **29**, 293.
4.   K. Okuno and S. Sakaguchi, *J. Hered.*, 1982, **73**, 467.
5.   Y. Sano, *Theor. Appl. Genet.*, 1984, **68**, 467.
6.   K. Denyer et. al., *Plant, Cell and Env.*, 1995, **18**, 1019.
7.   A.M. Smith, K. Denyer and C. Martin, *Plant Physiol.*, 1995, **107**, 673.
8.   R.B. Frydman and C.E. Cardini, *J. Biol. Chem.*, 1967, **242**, 312.
9.   F.D. Macdonald and J. Preiss, *Plant Physiol.*, 1985, **78**, 849.
10.   K. Denyer et. al., *The Plant J.*, 1993, **4**, 191.
11.   F.D. Macdonald and J. Preiss, *Plant Physiol.*, 1983, **73**, 175.
12.   A.M. Smith, *Planta*, 1990, **182**, 599.
13.   C.M. Hylton et. al., *Planta*, 1996, **198**, 230.
14.   C. Mu-Forster, Plant Physiol., 1996, in press.
15.   K. Denyer et. al., *Planta*, 1995, **196**, 256.
16.   S. Rahman, *Aust. J. Plant Physiol.*, 1995, **22**, 793.
17.   A. Edwards et. al., *The Plant J.*, 1995, **8**, 283.
18.   J. Marshall et. al., *The Plant Cell*, 1996, **8**, 000.
19.   K. Tomlinson, PhD Thesis, University of East Anglia, 1995.
20.   K. Denyer and A.M. Smith, *Planta*, 1992, **186**, 609.
21.   A. Edwards et. al., *Plant Physiol.*, 1996, **111**, 000.

22. J. Robyt, 'Enzymes in the synthesis and hydrolysis of starch.' In Starch: chemistry and technology, R.L. Whistler, J.N. BeMiller and E.F. Paschall, eds., Academic Press, Orlando, 1984, p. 87.
23. A.M. Smith and C. Martin, 'Starch biosynthesis and the potential for its manipulation.' In Biosynthesis and Manipulation of Plant Products: Plant Biotechnology Series, D. Grierson, ed., Blackie Academic and Professional, Glasgow, Scotland, 1993, Vol. 3, p.1.
24. C. Martin and A.M. Smith, *The Plant Cell*, 1995, **7**, 971.
25. T. Baba, M. Yoshii and K. Kainuma, *Stärke, 1987,* **39**, 52.
26. B. Delrue et. al., *J. Bacteriol.*, 1992, **174**, 3612.
27. J.L. Jane and J.F. Chen, *Cereal Chem.*, 1992, **69**, 60.

THE USE OF MUTANTS TO STUDY THE STRUCTURAL AND FUNCTIONAL PROPERTIES OF PEA STARCH

T. Ya. Bogracheva[a], S. Ring[b], V. Morris[b], J. R. Lloyd[c], T. L. Wang[c] and C. L. Hedley[c]

[a]ATO-DLO, Wageningen, The Netherlands.
[b]Institute of Food Research, Norwich, NR4 7UA
[c]John Innes Centre, Norwich, NR4 7UH

1 INTRODUCTION

Starch granules are semi-crystalline and consist of ordered regions containing double helices, formed by short amylopectin chains, and disordered or amorphous regions. In addition, the double helices are involved in the formation of the crystallites in starch granules. Starches are composed of two types of polymorph structures, A and B. These polymorphs differ in the packing of double helices into the crystallites; A type being more dense than B type. Maize starch is characterised by having A type polymorphs in its crystalline structure and the starch is called A-type. On the other hand, potato starch is characterised by having B type polymorphs in its crystalline structure and the starch is called B-type. Legumes in general, and peas in particular, are believed to be characterised by having both A and B polymorphs within their crystallites and legume starches are called C-type.[1,2] Heating starch suspensions in excess water results in disturbance of the ordered structures in the starch granules, a process known as gelatinisation.[3,4]

2 CRYSTALLINE STRUCTURE

We have used the range of starches from pea mutants described elsewhere (see Wang *et al* in this volume) to investigate the structure of C-type starch. The starch and amylose contents of the mutants used in the study are shown in Table 1. The starches were derived from lines which are effectively near-isogenic, which gives us a unique opportunity to study the effect of single gene mutations, known to affect starch content and composition, on the physico-chemical properties of the resulting starch.[5,6,7]

One of the most common methods for characterising the crystalline structure of starches is wide-angle X-ray diffraction. A-, B- and C-type starches are characterised by having different wide-angle diffraction patterns. The differences between the three types of starches are in the position and intensity of the peaks (Fig 1). The crystalline and amorphous parts of the starch contribute to these patterns.

Table 1 *Starch and amylose contents of the mutants used in the study*

Pea line	Seed shape	Starch (% dry weight)	Amylose (% starch)
Wild-type	round	50	35
r	wrinkled	36	65
rb	"	36	23
rug-3	"	12	12
rug-4	"	43	33
rug-5	"	29	43
lam	round	47	8

The total crystallinity of the starches was calculated by subtraction of the amorphous part using the following formula:-

$$Cr = \frac{\sum\limits_{2\theta^{\circ}} (I - I_{am})}{\sum\limits_{2\theta^{\circ}} I} \times 100$$

where: Cr - % total crystallinity; I - intensity of X-ray diffraction of starch; I_{am} - intensity of X-ray diffraction of amorphous part of starch and $2\theta^{\circ}$ - diffraction angle, changes from 4.5 to 28.0. It was found that the total crystallinity for starch from the wild-type and mutants ranged from 20 to 30%.

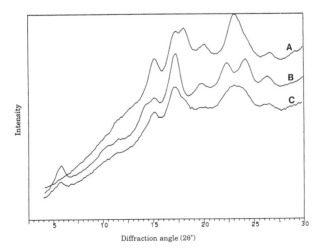

Figure 1 *X-ray diffraction patterns for A, B and C type starches.*

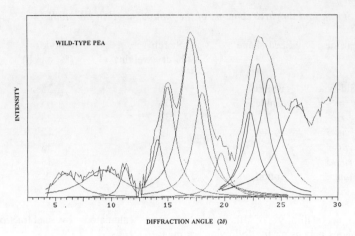

Figure 2 *Peak profile for wild-type pea starch obtained using a computer peak fitting
programme*

Subtraction of the amorphous part gives the X-ray pattern for the crystalline part of
the starch granule. The peak profiles of the crystalline parts of the starches (Fig 2), have
been derived using a computer peak fitting programme. This type of picture has been
obtained for A- and B-type starches, A and B spherulites and for all mutant pea starches.

The peak positions for A and B polymorphs are shown in Table 2. Starches from
both wild-type and mutant peas were shown to have the same peak positions and these
corresponded to the peak positions for the sum of A and B polymorphs. This showed
that all the pea starches were of C-type.

Table 2 *Peak positions profile for X-ray pattern in $2\theta°$*

Polymorphs			Wild-type and mutant pea starches
A	B	Sum of A&B	
-	5.8	5.8	6.1
9.8	9.8	9.8	9.4
11.2	11.2	11.2	11.2
-	14.0	14.0	14.1
15.0	14.9	15.0	15.0
16.9	17.0	16.9	16.9
17.9	-	17.9	18.0
19.9	19.6	19.7	19.7
-	22.2	22.2	22.1
22.9	-	22.9	22.9
23.9	24.0	23.9	23.9
26.4	26.4	26.4	26.4

Table 3 *Proportion of double helices and total crystallinity in different starches*

	Type of starch	Total crystallinity (%)	Double helices (%)
Wild-type pea	C	21	37
rb mutant pea	C	29	45
Potato	B	25	53
Maize*	A	39	42

* *data from S.Nara; Starch, 1978, 30, 183-186*

A method has been developed which uses the differences between the profiles for A and B polymorphs to determine the amounts of each in the C-type starches.[8] It was found that the amount of B type crystallinity ranged from about 40 to 90%.

The favoured method used to determine the relationship between the disordered and ordered structures in starch is solid-state NMR. It has been found that the peak at 81-83 ppm is characteristic for the NMR spectrum of amorphous starch and that there is virtually no signal in this region for crystalline materials.[9] The comparative relationships between the area of this peak and the total peak area for amorphous and native starches were used to define the proportions of ordered and disordered material.

Table 3 shows the total crystallinity and % double helical structures for starch from the wild-type and *rb* mutant, with potato and maize for comparison. In general, the proportion of double helical structures is greater than the proportion of total crystallinity within the starch granule. This was true for B- and C-type starches, but the difference, if any, was much less for A-type starch. Starch from wild-type peas had the lowest proportion of double helices of all the starches studied.

3 GELATINISATION

In aqueous suspensions at room temperature the amorphous phase of starch granules swells to some degree. During heating this swelling increases and becomes irreversible above certain temperatures. At the same time the cooperative melting of the crystalline structure, or gelatinisation, occurs. In starch suspensions two components, rigid and mobile, can be separated using NMR relaxation methods. The mobile component is related to the amorphous phase and the rigid component includes the ordered part of the starch. At room temperature the proportion relating to the rigid component is much greater than the proportion relating to double helices. This indicates, in agreement with the commonly accepted point of view,[10,11] that the rigid component consists of ordered structures coupled with parts of the amorphous material of starch granules. Consistent with the same accepted point of view, some interactions between the amorphous and ordered structures are disturbed during swelling and so the content of rigid component in starch granules becomes less. In this respect B- and C-type starches show similar behaviour, but this may differ from that of A-type starch.

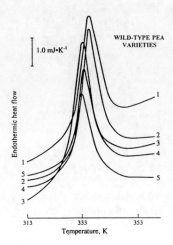

Figure 3 *DSC curves for starch from a range of pea varieties*

We used differential scanning colorimetry (DSC) for studying thermodynamic parameters of gelatinisation of pea starches. Quasi-equilibrium conditions on heating were used (an increase in temperature of 1°C/minute and starch concentrations of not more than 2%). Gelatinisation behaviour in excess water of starches from a number of wild-type round seeded peas is shown in Figure 3 and for starch from the wild-type relating to the mutants and for the *rb* mutant in Figure 4. All the starches had what appeared to be a single, or slightly double, narrow endothermic gelatinisation peak with some slight differences in peak temperatures.[5,12,13]

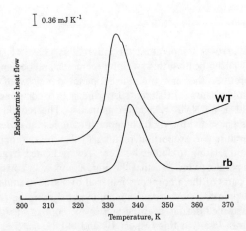

Figure 4 *DSC curves for starch from the wild-type and rb mutant line*

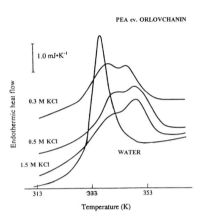

PEA cv. ORLOVCHANIN

1.0 mJ·K⁻¹

Endothermic heat flow

0.3 M KCl

0.5 M KCl

WATER

1.5 M KCl

313   333   353

Temperature (K)

Figure 5 *Effect of KCl on the gelatinisation of starch from a wild-type pea cultivar*

During gelatinisation water acts like a plasticiser. Any changes which alter the quality of the water, such as the addition of salts, therefore, should affect the thermodynamic parameters of gelatinisation. Figure 5 shows the effect of different KCl concentrations on gelatinisation of wild-type pea starch. The addition of the salt changes the single peak of transition in excess water into a double peak. The starch from the wild-type pea relating to the mutants and from the *rb* mutant showed similar behaviour in salt solutions (Fig 6).

The behaviour of the two peaks under different concentrations of salt was similar to that of A-type (e.g. maize) and B-type (e.g. potato) starches under the same conditions.[14] This indicates that there are two different independent cooperative transitions during gelatinisation of pea starches; one corresponding to the melting of A polymorphs (higher temperature peak) and the other to B polymorphs (lower temperature peak).[5,12,13,14] This hypothesis was supported by an experiment in which wild-type pea starch was heated in 0.6M KCl solution until just after the first transition and then rapidly cooled and then reheated. During the second heating there was only one transition peak which occurred at the higher temperature. Using wide-angle X-ray crystallography, the peak remaining after the first heating of the starch was then shown to have attained the characteristics of A-type starch.

With regard to gelatinisation behaviour in excess of water, the mutant starches could be divided into two groups. Starches from the first group (wild-type, *rb*, *rug-3*, *rug-4* and *lam* lines) all showed a narrow cooperative endothermic peak of gelatinisation, which is representative of a first order transition. There was a 10°C range of peak temperatures for the starches within this group, the lowest peak temperature being for the *lam* mutant and the highest for *rug-3*. Starches from the second group (*r* and *rug-5* lines) showed a very small wide peak, or no peak, during gelatinisation, indicating that these transitions cannot be referred to as first order. From an analysis of the thermodynamic parameters of transition for the first group of starches there was no evidence of a direct relationship between the amylose content, the content of A and B polymorphs and the thermodynamic parameters of their gelatinisation.

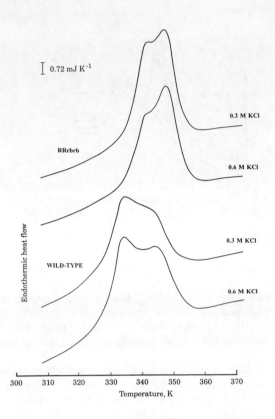

Figure 6 *Effect of KCl on gelatinisation of starch from wild-type and rb mutant lines*

Differences between the two groups of starches with respect to gelatinisation could also be seen in the behaviour of their rigid and mobile components using NMR relaxation. During heating, using similar conditions to those in the DSC experiment, the rigid components of the second group of starches were reduced compared with the first group and, unlike the first group, disappeared to zero after about 40°C. This indicated that the process of disruption of the ordered structures during heating was different between the two groups.

The changes occurring during gelatinisation are related to the changes in the ordered structures and involve the crystalline part of the amylopectin. This was investigated using ion exchange chromatography. The branching patterns for the crystalline parts of maize and potato starch, as well as for the starches from the first group of pea lines all showed single peak curves. The degree of polymerisation (DP) of the crystalline parts of amylopectin for the second group of pea lines, however, showed multipeak curves with a large proportion of the chains having a higher DP. This indicates that the crystalline part of the amylopectin is constructed differently for the two groups of pea lines.

4 CONCLUSIONS

The unique collection of pea mutants which we have identified are ideal for studying the effects of specific mutations on the structure and function of starch. The differences in polymorph content found in the mutants cannot account for all of the properties of C-type starches. Although the wild-type and mutant pea starches are of C-type and contain A and B polymorphs the properties of these starches are not intermediate between those of A- and B-type starches. A wide range of gelatinisation behaviour was found between the starches from the different pea lines.

5 ACKNOWLEDGEMENTS

We would like to thank Paul Cairns, Lorraine Barber and Natalia Davydova for providing some of the experimental results.

References

1. A. Sarco and H-CH. Wu, *Starch/Starke*, 1978, **30**, 73.
2. A. Imberty and S. Perez , *Biopolymers*, 1988, **27**, 1205.
3. Y. M. V. Blanshard, in: Starch: Properties and Potential, ed. T.Galliard. Published for Society of Chemical Industry by Lohn Wiley, Chichester, 1987, 16.
4. D. Cooke, M. J. Gidley, *Carbohydr. Res.*, 1992, **227**, 103.
5. C. L. Hedley, T. Ya. Bogracheva, J. R. Lloyd, T. L. Wang, in: Agri-Food Quality, ed. G. R. Fenwick, C. L. Hedley, R. L. Richards, S. Khokhar, 1996 (in press).
6. T. L. Wang, C. L. Hedley, *Pisum Genet.*, 1993, **25**, 64.
7. K. Denyer, L. Barber, R. Burton, C. L. Hedley, C. M. Hylton, S. Johnson, D. A. Jones, J. Marshall, A. M. Smith, H. Tatge, K. Tomlinson and T. L. Wang, *Plant, Cell and Environ.*, 1995, **18**, 1019.
8. P. Cairns, T. Bogracheva, C. L. Hedley, V. Morris, S. Ring., *Carbohydr. Polym.* (Submitted for publication).
9. M. J. Gidley and S. M. Bociek, *J. Am. Chem. Soc.*, 1995, **107**, 7040
10. C. G. Biliaderis, C. M. Page, T. J. Maurice and B. O. Juliano, *J. Agric. Food Chem.*, 1986, **34**, 6.
11. I. J. Colquhoun, R. Parker, S. G. Ring, L. Sun and H. R. Tang, *Carbohydr. Polym.*, 1995, **27**, 255.
12. N. I. Davydova, S. P. Leont'ev, Ya. V. Genin, A.Yu. Sasov and T. Ya. Bogracheva, *Carbohydr. Polym.*, 1995, **27**, 109.
13. T. Ya. Bogracheva, N. I. Davydova, Ya. V. Genin and C. L. Hedley, *J. Exp. Botany*, 1995, **46**, 1905.
14. T. Ya. Bogracheva, S. P. Leontiev, Ya. V. Genin, *Carbohydr. Polym.*, 1994, **25**, 227.

CHARACTERIZATION OF STARCH FROM GENETICALLY MODIFIED POTATO AFTER TRANSFORMATION WITH THE BACTERIAL BRANCHING ENZYME OF *ANACYSTIS NIDULANS.*

A. J. Kortstee, L. C. J. M. Suurs, A. M. G. Vermeesch and R. G. F. Visser

Graduate School of Experimental Plant Sciences
Department of Plant Breeding
Agricultural University Wageningen
P.O.Box 386, 6700 AJ, Wageningen, NL.

1 INTRODUCTION

Starch is the most abundant storage carbohydrate found in higher plants. Starch consists of two types of glucose polymers; amylose an essentially unbranched α-1,4 linked glucose polymer and amylopectin, consisting of α-1,4 linked chains with α-1,6 branchpoints. Potato tuber starch consists of about 20 to 25 % of amylose. The enzyme responsible for the formation of amylose is Granule Bound Starch Synthase (GBSS); amylopectin is synthesized by the combined actions of the enzymes Soluble Starch Synthase (SSS) and Branching Enzyme (BE) [1].

Suppressing the expression of one or more of the starch synthesizing genes by means of antisense RNA technology results in starches with an altered composition as was reported before [2-4]. Partial inhibition of the GBSS gene expression results in starch with a lowered amylose content. The reduced amount of amylose was found in a restricted zone in the granule in a core of varying size at the hilum of each granule [5]. In our attempt to influence starch branching degree we introduced bacterial branching enzyme genes in potato. The glgB gene of *A.nidulans* was placed under the transcriptional control of the potato GBSS gene and transitpeptide sequence including box1, the substrate binding site, of the mature protein. Expression of this construct in potato tubers lead in some transformants to an increased branching degree of the starch. In one of the transformants, number AN-24, also a lowered amylose content was found. The question arose whether the decreased amylose content was caused by the expression of the bacterial branching enzyme or whether it was a side effect of the construct we used and if apart from the lowered amylose content the amylopectin structure had changed as well. We decided to investigate some properties of this type of starch and to compare it to starch with a lowered amylose content as a result of antisense expression of the GBSS gene and to starch with an increased branching degree as a result of the expression of the heterologous branching enzyme gene of *A.nidulans*. By comparing physico-chemical properties of AN-24 starch with those two types of starches it may be possible to determine whether AN-24 has, apart from a lowered amylose content, an altered amylopectin structure.

2 RESULTS

2.1 Plant material

Tetraploid potato clone K892002 (2002) (Karna, Valthermond, Holland) transformed with construct pKGBA50 or pGB50 [6] carrying the GBSS cDNA of potato in anti-sense orientation under the control of the potato GBSS promoter or the CaMV promoter. Individual transformants, selected for incomplete GBSS inhibition, used in this study were: tBK50-10, tBK50-34 and tB50-42. Diploid potato clone A16 was transformed with construct $pB_{19}13AN$, carrying the glgB gene of *A.nidulans* under the transcriptional control of the potato GBSS promoter sequence including the transitpeptide sequence and box1, the substrate binding site, of the mature protein. Individual selected clones were AN-9, AN-14 and AN-24.

2.2 GBSS analysis

Transformants expressing the pKGBA50 construct were found to have a decreased expression of the GBSS mRNA and GBSS protein (data not shown). Enzymatic activity of the GBSS protein had also decreased as can be seen in Table 1. The levels of expression of the endogenous branching enzyme had not changed. Tubers of plants transformed with the $pB_{19}13AN$ construct were found to have normal GBSS mRNA and protein levels. One of the transformants, number AN-24, showed a lowered GBSS activity. The levels of endogenous branching enzyme mRNA and protein had not changed (data not shown).

Table 1 *GBSS enzyme activity and apparent amylose percentage*

Clone	GBSS activity* pmol/min/mg starch	Amylose content* %
K892002	100 ± 10	20 ± 1
tBK50-10	45 ± 30	9 ± 3
tBK50-34	45 ± 30	12 ± 7
tB50-42	45 ± 30	15 ± 4
A16	50 ± 2	22 ± 1
AN-9	57 ± 3	23 ± 1
AN-14	55 ± 3	21 ± 2
AN-24	33 ± 3	13 ± 2

* All measurements are the average of three independent measurements

2.3 Starch granule size and morphology

Granule size and morphology were determined by the Coulter Counter and by microscopical analysis of iodine stained starch. Granule size and shape did not seem to have been altered for the transformed plants compared to their controls. Looking at iodine stained starch it could be observed that the anti-sense inhibited GBSS plants had red staining starch granules with a blue core of varying size, from very small to granule filling as was reported before [5]. The untransformed control contained completely blue staining granules. Transformants with the pB1913AN construct all contained completely blue staining starch granules except for transformant AN-24 which contained red staining granules with a blue core, varying in size from very small to completely granule filling (Fig. 1).

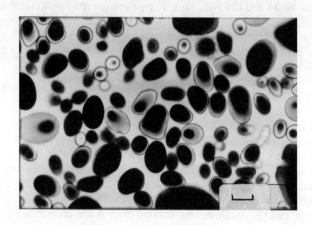

Figure 1 *Starch granules from transformant AN-24 stained with iodine, the bar denotes 20 μM*

2.4 Starch composition and structure

The apparent amylose content of the transgenic starches was determined by a spectrophotometric method [7]. Untransformed potato (=control) starch of clone 2002 contained about 20% amylose, the GBSS inhibited transformants showed a decreased amylose content of about 9-15% (Table 1). Transformants with the $pB_{19}13AN$ construct showed no change in amylose content except for transformant AN-24 which had a lowered amylose content of 13 %. The untransformed control A16 had an apparent amylose content of 22 %.

The λmax and Blue Value of solubilized starch were determined spectrophotometrically. The λmax of starches with a lowered amylose content is lower like the Blue Value. The branching degree of the starch was determined by the Luff-Schoorl method [12]. The Dextrose Equivalent (DE) after isoamylase digestion is expressed as the percentage of α-1,6 branchpoints per dry weight starch. The DE in the antisense transformants has increased as a result of the lowered ratio amylose to amylopectin. As amylose contains nearly no branchpoints compared to amylopectin the percentage of branchpoints per dry weight will increase if the amylose content is lowered. In the transformants expressing the glgB gene of *A.nidulans* the amylose content had remained the same, but the DE had increased. The amylopectin of these starches contained more short chains, the so called A chains in their amylopectin (data not shown). For transformant AN-24 a lowered amylose content was found and a higher DE, but also an increased ratio of short to longer chains after HPLC analysis of isoamylase digested amylopectin (data not shown). The increased DE after isoamylase digestion of AN-24 starch was caused by two different things; the lowered amylose content as well as the presence of more short chains in the amylopectin.

Table 2   *λmax, Blue Value and branching degree (in DE)*

Clone	*λmax (nm)*	*Blue Value (at 680 nm)*	*DE % α - 1 , 6 branchpoints*
K892002	578	1.54	4.0
tBK50-10	558	0.99	4.7
tB50-42	570	n.d	4.7
A16	574	1.2	3.6
AN-9	576	1.15	4.1
AN-14	575	1.18	4.5
AN-24	567	0.65	4.5

n.d = not determined

2.5 CL2B chromatography

Amylopectin and amylose were separated by size exclusion chromatography with a CL2B column. The fractions eluted from the column are spectrophotometrically scanned after iodine staining. Amylopectin is eluted from the column first as a large narrow peak, sometimes with a shoulder. The λmax of fractions of this peak after iodine complexation was approximately 560 nm, confirming the identity of the eluted glucan. Immediately after the amylopectin a glucan is eluted with a λmax of ≥ 600 nm, indicating amylose as a smaller but broader peak. The elution profiles presented in Figure 2. of the different starches are very similar except that it is clear that starch from tBK50-10 and AN-24 contain less amylose.

K89.2002

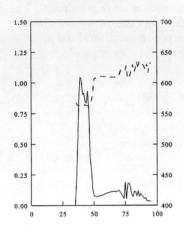

TBK50-10

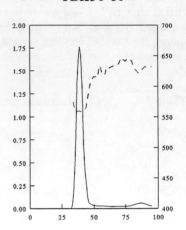

A-16

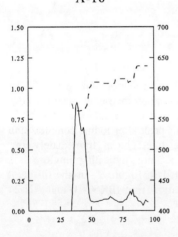

AN-24

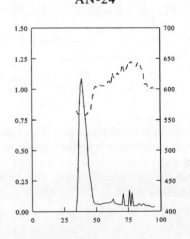

Figure 2 *CL2B profile of starches,*
——— = *Absorbance at λmax (left Y-axis),*
------ = *λmax (right Y-axis),*
the x-axis represents the fraction number.

2.6 Swelling power

The swelling power of starch in water was determined during heating and is presented in Fig. 3. Normal amylose containing potato starch swells linear in time with increasing heat. Amylose-free starches swell much faster but usually begin to swell at a higher temperature. The starches of plants expressing the antisense GBSS gene showed a faster swelling pattern than the untransformed control. Solubility at 75 °C was lowered from 8% to 1-2% for the GBSS inhibited starches which was also observed for amylose-free starch (data not shown). Starch from plants expressing the *A.nidulans* glgB gene showed a lowered swelling power compared to the untransformed control. The transformant with the lowered amylose percentage, AN-24, showed the characteristic lowered solubility (2% instead of 8%) as well as a lowered swelling power compared to the control.

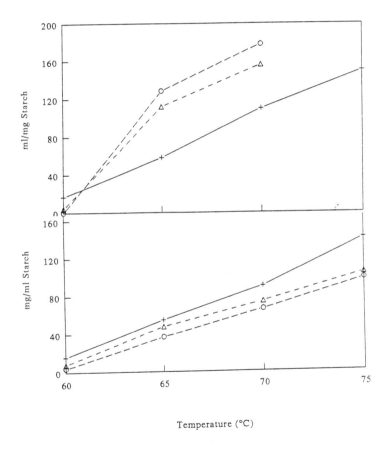

Figure 3 *Swelling power of starch in water. Swelling power is determined as the volume of water taken up by 1 gram of starch. Top: + = K892002, Δ = tB50-42, o = tBK50-10, bottom: + = A16 control, Δ = AN-14, o = AN-24.*

2.7 Thermal analysis

2.7.1 DSC. The temperature of onset of gelatinization (To) and other gelatinization characteristics were measured by DSC. The results of the DSC analysis are presented in Table 3. Starch of transformants AN-14 and AN-24 showed a slight increase in the temperature of onset of gelatinization (To) compared to the untransformed control (+ 1 °C). Likewise the peak temperature (Tp) had shifted to a higher temperature, but similar enthalpie (ΔH) and gelatinization range (Tp-To) were found. Starches from tubers with a lowered amylose content showed an increase in the temperature of onset of gelatinization (To) of about 3 °C. Likewise the Tp had shifted to a higher temperature and the enthalpy had also increased.

Table 3   *Thermal analysis by DSC of transgenic starches*

Clone	To (°C)	Tp (°C)	ΔH (J/g)
KG892002	60.3	64.5	16.5
tBK50-10	64.0	68.8	21.7
tBK50-34	62.9	67.2	24.2
tB50-42	63.5	69.5	21.3
A16-gus	59.3	63.3	16.7
AN-9	59.7	64.2	19.4
AN-14	60.8	65.5	21.1
AN-24	60.9	66.6	18.4

2.7.2 Bohlin. Dynamic rheological properties were determined by applying a small oscillilating shear deformation using a Bohlin VOR Rheometer as described in literature [8]. The results are shown in Fig. 4. The storage modulus (G') of the starch/water suspensions increased sharply upon heating. After a maximum was reached the viscosity decreased during further heating. After cooling the viscosity increased again as a result of retrogradation of, presumably, amylose. The difference between the control 2002 and the transformant with partial GBSS inhibition and a lowered amylose content tBK50-42 can be seen from Fig 4. The viscosity of the transformant, the onset of gelatinization, starts to increase at a higher temperature compared to the untransformed control. The maximum viscosity however is reached at the same time/temperature and of similar/identical height. The viscosity as a result of retrogradation after cooling is almost absent in the antisense GBSS transformant with the lowered amylose content. Starch from transgenic plants expressing the glgB gene of *A.nidulans* showed the same viscosity profile as the untransformed control except for a lowered peak viscosity for the starch with an increased branching degree (data not shown). Starch from transformant AN-24 with the lowered amylose content not only had a lowered peak viscosity but also showed a higher temperature of onset of gelatinization. The increase in viscosity as a result of retrogradation of amylose was observed for those starches.

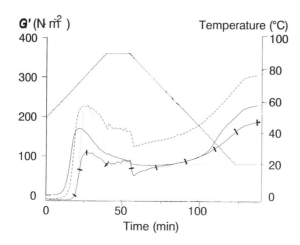

Figure 4 *Changes in the storage moduli (G') of 5 % (w/v) starch suspensions against time and temperature, --- = K892002, ... = tB50-42, ———— = A16, ——+—— = AN-24.*

3 DISCUSSION

In this study we compared some physico-chemical properties of starch from transgenic potato plants. Plants transformed with a construct carrying a bacterial branching enzyme showed an increased branching degree of the starch. One of the transformants, number AN-24, had a lowered amylose content. To determine whether the effect on the starch in AN-24 was only on the amylose or also on the amylopectin fraction of the starch, starch of GBSS inhibited plants (containing a decreased amount of amylose) was also analysed and compared to starch from the glgB expressing plants. All transformants contained starch granules of normal size and morphology. Microscopic analysis of iodine stained starch granules showed the presence of amylose in a restricted area of the starch granules for transformant AN-24 and the antisense GBSS expressing plants. The reduced amount of amylose was present in a core around the hilum of the granule. For plants with a reduced amount of amylose also a reduction of the GBSS activity could be observed as was reported before [5]. Starch with a lowered amount of amylose had, when dissolved, a lower λmax and Blue Value compared to the normal amylose containing starches, as was expected from the established relationship between amylose content and λmax and Blue Value [9]. The DE after isoamylase digestion was found to be higher in starch with a lowered amount of amylose on account of relative higher amount of amylopectin per dry weight starch. The DE (after isoamylase digestion) of starches with a normal amylose content, but expressing the glgB gene of *A. nidulans* had also increased. This however was the result of the presence of more short chains in the amylopectin. Gel permeation chromatography was used to separate the amylopectin and amylose by weight/volume.

No differences were found between control starch and starches from transgenic plants except for the amount of amylose. Not only structural, but also swelling and gelation properties of the starches were investigated. The swelling power measured during heating of all the starches from plants expressing the glgB gene were essentially the same, including transformant AN-24. The solubility at 75 °C was about half for starch from transformant AN-24 compared to the others. The other starches (from plants with inhibited GBSS expression) with a decreased amylose content showed a different swelling power from the controls. These starches showed very fast and almost unrestricted swelling, comparable to amylose-free mutant starch. The solubility of the starches with a lowered amylose content was much lower compared to the controls. Thermal analysis by DSC showed the onset of gelatinization of starch in water. Starch from antisense inhibited GBSS plants showed an increase of temperature of onset of gelatinization of almost three degrees Celcius compared to the untransformed amylose containing control. The influence of the presence of amylose on the temperature of the onset of gelatinization (To) was shown before [10] and is apparently not only a property of the amylopectin as is widely believed. Starch from transformant AN-24 had an increased temperature of onset of gelatinization, but not so much as the antisense starches. Transformant AN-14 with the highest branching degree of the amylopectin showed an increase in the temperature of onset of gelatinization compared to the untransformed control. This could be the result of an altered crystallinity because of tighter stacking of the more branched amylopectin or more entanglement between the more branched amylopectin molecules.

The dynamic rheological properties were determined by Bohlin. Starch from the antisense GBSS expressing transformants showed gelatinization at a higher temperature, maximum viscosity was reached at the same temperature but an increase in G' after cooling as a result of retrogradation was almost absent. The increase in G' for starch from transformant AN-24 was just as the starch from antisense transformants at a higher temperature compared to the control. However, the peak viscosity was lower for transformant AN-24 compared to the control. This decrease of the peak viscosity could also be seen for transformants with a normal amylose content but a higher branching degree of the amylopectin. Afer cooling an increase in the storage modulus was similar for all the glgB carrying transformants.

Starch from transformant AN-24 seemed to express some physico-chemical properties typically related to the amylose content such as the change in λmax and B.V., the increased temperature of onset of gelatinization as could be seen from the DSC and Bohlin analysis and the lowered solubility in water. Other characteristics of AN-24 starch could not simply be attributed to the decreased amount of amylose. These characters, including the unique gel permeation chromatogram and the decrease in swelling power as well as the decreased peak viscosity seem to be related to structural changes in the amylopectin. They are similar to the changes in properties seen for transformants with a normal amylose content and an increased branching degree of the amylopectin. So, probably, transformant AN-24 has an altered amylopectin structure as the result of the expression of the bacterial branching enzyme. The lowered amylose content could be the result of the action of the heterologous branching enzyme, synthesizing amylopectin-like material from amylose. Another possible explanation for the lowered amylose content could be that by using the potato GBSS promoter and transitpeptide sequence including a substantial part of the mature protein causes so called co-suppression. Co-suppression can cause the inhibition of gene expression of both the inserted and the endogeneous GBSS gene by an unknown mechanism [11].

4 ACKNOWLEDGEMENTS

We would like to thank T. van de Brink and J. Rijksen for taking care of the plants in the greenhouse, Dr. A. G. J. Kuipers for the gift of the antisense GBSS starches, Dr. C.J. A. M. Keetels for the Bohlin analysis and helpfull discussions and Nik Boer from Perkin & Elmer (Holland) for helping with the DSC analysis.

References

1. C. Martin and A.M. Smith, *The Plant Cell*, 1995, **7**, 971.

2. B. Müller-Rober, U. Sonnewald and L. Willmitzer, *EMBO Journal*, 1992, **11**, 1229.

3. R. G. F. Visser, I. Somhorst, G. J. Kuipers, N. J. Ruys, W. J. Feenstra and E. Jacobsen, *Mol.Gen.Genet.*, 1991, **225**, 289.

4. E. Flipse, L. Suurs, C. J. A. M. Keetels, J. Kossmann, E. Jacobsen and R. G. F. Visser, *Planta*, 1996, **198**, 340.

5. A. G. J. Kuipers, E. Jacobsen and R. G. F. Visser, *The Plant Cell*, 1994, **6**, 43.

6. A. G. J. Kuipers, W. J. J. Soppe, E. Jacobsen and R. G. F. Visser, *Mol.Gen.Genet.*, 1995, **246**, 745.

7. J. H. M. Hovenkamp-Hermelink, J. N. De Vries, P. Adamse, E. Jacobsen, B. Witholt and W. J. Feenstra, *Potato Research*, 1989, **31**, 241.

8. C. J. A. M. Keetels, PhD Thesis, University of Wageningen, 1995.

9. Y. J. Wang, P. White, and L. Pollak, *Cereal Chem.*, 1993, **70**, 199.

10. E. Flipse, C. J. A. M. Keetels, E. Jacobsen and R. G. F. Visser, *Theor.Appl.Gen.*, 1996, **92**, 121.

11. E. Flipse, I. Straatman-Engelen, G. J. Kuipers, E Jacobsen and R. G. F. Visser, *Plant.Mol.Biol.*, **31**, 731.

12. A. J. Kortstee, A. M. G. Vermeesch, B. J. de Vries, E. Jacobsen and R. G. F. Visser, 1996, *The Plant Journal*, **10**, 83.

ACKNOWLEDGEMENTS

We would like to thank Dr. Brian J. H. Smith for the assistance of the plants in the greenhouse. Dr. S. C. Angus for the gift of the antisera (GUS antibody). Dr. P. J. A. M. Kaanlic for the Gaussian technical and helpful suggestions and Ms. Barbara van Loon and Mrs. Lisbet Holland/Harp-Chen with the BUS analysis.

REFERENCES

1. C. Martin and A.W. Smith, The Plant Cell, 1995, 7, 971.

2. W. Martindale, J. Sambrook and J.P. Whitlock, PNAS J. Biol. 1991, 47, 1929.

3. G.C. Vrede, J. Benson-van Schagen, V.L. 1992, and J. Benson-van Schagen, Mol. Gen. Genet., 1991, 225, 289.

4. J. Jansen, J. Ghij, F.C. V. Goers, J. Kesteren, B. Bleeker and R.G.F. Visser, Plant Cell, 1992, 90.

5. J.W. Jansen, H. van Loon and R.H. V. Visser, The Plant Cell, 1994, 76, 489.

6. C.F. Bauer, W.F.J. Semp, J. Jacobsen and B.C.E. Weber, Mol. Gen. Genet., 1993, 236, 243.

7. H. Sm. Beveridge-Heine, J. de la Vrede, R. Adams, F. Jacobsen, E. Witte, I. and W.J. Benson, Transgenic Res., 1999, 31, 2418.

8. G.C. A. Kesler, PhD thesis, University of Wageningen, 1999.

9. A.E. Watts, H. Visser and C. Friend, Cereal Chem., 1997, 70, 169.

10. C. Wu, J. McGrath, L. Jacobsen and R.G.F. Visser, Theor. Appl. Gen., 1996, 92, 274.

11. B. Tupta, L. Ernaning, Jegels, G.D. Kuiper, J. Jacobsen and R.G.F. Visser, Transgenic Res., 21, 793.

12. A.J. Kortsee, A.M.G.G. Vermeesch, B.J. de Vries, H. Jacobsen and R.G.E. Visser, 1996, The Plant Journal, 68, 83.

Poster Presentations

EFFECT OF SUGARS ON THE THERMAL AND RETROGRADATION PROPERTIES OF OAT STARCHES

R. Hoover and N. Senanayake
Department of Biochemistry, Memorial University of Newfoundland St John's, Newfoundland A1B 3X9 Canada

The thermal and retrogradation properties of oat starches from two cultivars (NO 753- 2 and AC Stewart) were characterized in the presence of glucose, fructose and sucrose at a concentration of 36% w/v. In both oat starches, amylose leaching (AML) and swelling factor (SF) decreased in the presence of sugars (sucrose > glucose > fructose). These decreases were more pronounced in AC Stewart starch, The decrease in AML, showed that sugars interact with amylose chains within the amorphous regions of the starch granule. The gelatinization transition temperature and the enthalpy of gelatinization increased in the presence of sugars (sucrose > glucose > fructose), The above increase was also more pronounced in AC Stewart starch. The decrease in SF and the increase in gelatinization parameters indicated that these changes were influenced by the interplay of factors such as, starch-sugar interaction, changes in water structure in the presence of sugars, and the antiplastisizing properties of sugars relative to water. The retrogradation enthalpy and the X- ray diffraction intensities of NO 753-2 and AC Stewart starch gels (stored at 4°C) increased in the presence of sugars (glucose > fructose > sucrose). These changes were more pronounced in NO 753-2 starch. The results showed that interaction (during storage) between leached amylopectin and sugar molecules was the main causative factor influencing oat starch retrogradation.

THE PHASE BEHAVIOUR OF STARCH IN TERNARY AQUEOUS SOLUTIONS

G.K. Moates, R. Parker, and S.G. Ring
Food Molecular Biochemistry Department, Institute of Food Research,
Norwich Research Park, Colney Lane, Norwich, NR4 7UA.

The interactions of starch with water and low molecular weight solutes have been studied using differential scanning calorimetry (DSC), equilibrium dialysis and precision densitometry, sorption equilibria and mixing calorimetry. The results are analysed using thermodynamic approaches and Flory-Huggins theory. The increase in the melting temperature of B-type spherulitic short-chain amylose on the addition of low molecular weight carbohydrates can be interpreted simply as an effect of their molecular size. Equilibrium dialysis and precision densitometry shows that there is a weak interaction between carbohydrates (sorbitol, glucose, glycerol) and amylopectin but a strong preferential interaction (compared with water) with urea and guanidinium thiocyanate (GTC). The strong interaction of urea and GTC results in a depression of the melting temperature of B- type spherulites. Water sorption equilibria and mixing calorimetry provide further information on the thermodynamics of the interactions of water with starch and starch hydrolysis products. The implications of these results for the understanding of the processing and functionality of starch in food and non-food applications is discussed.

ENTHALPY RELAXATION IN GLASSY AMYLOPECTIN

S.J. Livings[1], C. Breach[1], A.M. Donald[1] and A.C. Smith[2]
[1]Cavendish Laboratory,University of Cambridge, Madingley Road,
  Cambridge CB3 OHE, UK
[2]Institute of Food Research, Norwich Laboratory, Colney, Norwich NR4 7UA, UK

The effect of ageing starch based systems below their glass transition temperature has been studied by DSC. The appearance of a peak below the T_g has been monitored in amylopectin samples of controlled water content. Both the position and precise temperature of this sub-T_g peak has been followed as a function of the sample's aging history - specifically changing both the time and temperature of storage. Analysis of the time development of the enthalpy change and peak position was carried out. It was found that both parameters varied logarithmically with annealing time for a fixed temperature of annealing. Additionally there was a linear relation between the enthalpy change and the difference between the annealing temperature and the T_g. These findings demonstrate that the origin of the sub-T_g peak lies in enthalpy relaxation. The data can be fitted within those models developed to explain enthalpy relaxation in synthetic glassy polymers to yield structural parameters for glassy amylopectin.

THE GLASS TRANSITION AND SUB-T_G ENDOTHERM OF AMORPHOUS AND NATIVE POTATO STARCH AT LOW MOISTURE CONTENT

H.J. Thiewes and P.A.M. Steeneken.
Netherlands Institute for Carbohydrate Research TNO, 9723 CC Groningen, The Netherlands

The glass transition temperature (T_g) marks the transition in mechanical behaviour from brittle to ductile in amorphous and semicrystalline polymers. Modification of polymer properties by physical treatment can be effected only above T_g.

The T_g of native starch remains a matter of much debate because of its strong dependence on the moisture content and the masking of the glass transition in DSC by other thermal events.

We have attempted to locate T_g in native potato starch at 16% moisture content by making use of the phenomena of annealing and enthalpy relaxation. The latter phenomenon relates to thermoreversible physical aging, which occurs only below T_g, whereas annealing (perfection of crystalline organization) is possible only above T_g.

By means of a series of aging experiments at -21 - 70°C, which is below the presumed T_g we were able to show that the position and magnitude of the so- called sub-T_g endotherm were in agreement with the relationships for enthalpy relaxation outlined by Hodge and Berens. There was no evidence for annealing to occur in the temperature range of these aging experiments. Therefore, the sub-T_g endotherm is due to enthalpy relaxation rather than to carbohydrate-water interactions. This endotherm reflects the thermal history of the starch sample.

After erasing the thermal history by very short heating at a temperature slightly above the presumed T_g a distinct glass transition with an onset at 80°C could be observed in native potato starch. Prolonged thermal treatment at 95°C (somewhat higher than the presumed T_g), resulted in a narrowing of the first melting endotherm, which is evidence for annealing.

T_g-onset of amorphous potato starch at 16% moisture content was observed at 4 °C, in agreement with literature data. The difference in T_g between native and amorphous starch is probably due to motional restrictions of the amorphous chain segments by crystalline domains and a higher concentration of physical crosslinks in the native starch granules.

INGRESS OF WATER INTO GLASSY AMYLOSE PELLETS

I. Hopkinson and R.A.L. Jones
Polymers and Colloids Group, Cavendish Laboratory, Cambridge. CB4 0HE

The ingress of water vapour into glassy pellets of amylose has been studied using stray field nuclear magnetic resonance imaging (STRAFI) to produce one dimensional water concentration profiles as a function of depth, from the surface where sorption occurs, and time. The spatial resolution for this technique is around 50µm. The observed concentration profiles are characteristic of a system where the mutual diffusion coefficient, D, is a function of the concentration, c, of the water penetrant. The concentration of water in the surface layer of the amylose increases over a period of 24 hours. This behaviour is explained in terms of limitation by mass transport in the water vapour. The form of D(c) has been obtained by use of the Boltzmann transform and numerical models, incorporating a water vapour transport term and the general Fickian diffusion equation. The behaviour observed is not characteristic of Case II diffusion.

STRUCTURE PROPERTY RELATIONSHIP IN TROPICAL TUBER STARCHES

S.N. Moorthy
Crop Utilisation and Biotechnology Division, Central Tuber Crops Research Institute, Trivandrum-695 017, India

The tropical tuber starches possess widely varying physicochemical and functional properties among themselves. An attempt has been made to correlate the structural features of these starches with their functional properties. XRD, DSC, Viscometry, Laser Raman Spectroscopy, [13]CCP/MAS NMR etc. have been used to characterise the starches. It was seen that the yam starches have totally different functional properties compared to the aroid starches. Cassava and sweet potato starches possess their own characteristics. The XRD pattern of all the yam starches was B and that of all aroid starches A. The pasting temperature of the aroids was much lower by viscometry and DSC. The viscosity patterns and stability were also vastly different. There was no correlation between the starch granule size and amylose contents with the rheological properties. Feeding trials on rats showed the digestibility was higher for raw aroid starches and low for the yam starches. However cooking brought the digestibility of both the starches to almost equal levels. The results with DSC, CP-MAS NMR and Laser Raman spectroscopy also revealed minor variations among the starches and the possible implications are discussed in the paper.

VARIATIONS IN STARCH COMPOSITION AND PROPERTIES IN BARLEY CULTIVARS WITH RELATION TO MALTING QUALITY

C.S. Brennan[1], N. Harris[1], A.S. Tatham[2], P.R. Shewry[2], I. Cantrell[3], D. Griggs[3] and D. Smith[4].
[1]Department of Biological Sciences, University of Durham, South Road, Durham, DH1 3LE.
[2]Long Ashton Research Station, Long Ashton, Bristol. BS18 9AF.
[3]Pauls Malt Ltd, Bury St Edmonds, Suffolk, IP32 7AD.
[4]Plant Breeding International Maris Lane, Trumpington, Cambridge, CB2 2LQ.

The structural and physico-chemical properties of starch granules from a range of barley cultivars exhibiting differing malting qualities were examined. Both microscopical and biochemical techniques were employed. Starch granules within the endosperm of good- malting cultivars were observed to have far less Starch Associated Proteins (SAP's) than were observed associated with starch from poor-malting endosperms. SEM images of malted endosperms demonstrated a gradual progression of starch degradation in good- malting cultivars during malting. In contrast, starch granules of poor-malting cultivars showed retarded degradation with a high retention of SAP's at the end of malting. It is proposed that the SAP's form an effective barrier around the starch granule, rendering some granules less susceptible to degradation by amylolytic enzymes. Differences in the binding of SAP's to starch granules affects the eventual malting performance of the grain. The relationship of variations in SAP may be related to differences in the amylose/amylopectin ratio in starch.

DISSOCIATION OF AMYLOSE-LIPID COMPLEXES IN BARLEY STARCH DISPERSIONS

P. Forssell, S. Shamekh and K. Poutanen
VTT Biotechnology and Food Research, P.O. Box 1500, FIN- 02400, VTT Finland

The lipids in isolated starches are mainly lysophospholipids, which are assumed to exist as amylose-lipid inclusion complexes in native starch granules. The amylose-lipid complexes (AML) will dissociate when starch dispersions are heated to about 100°C and the complexes reorganize during cooling the dispersion to room temperature. The addition of polar lipids such as lysophospholipids, are known to retard crystallization of starch gels. The mechanism of the effects of polar lipids on starch gels is not known but it is linked to their ability to form complexes with the amylose. However, the lysophospholipids retard starch gel crystallization also when added in excess, which means more than is needed to complex the amylose. This behaviour has been explained as being caused by the formation of complexes between the lipids and the amylopectin.

We studied the dissociation of the AML-complexes in dilute barley starch dispersions as a means of elucidating especially the way the added lipids interact with the starch polymers. The native starch dispersion (1%) as well as the starch dispersion in which lysophospholipids had been added, was heated up to 85-100°C and centrifuged. The lipid and the amylose contents of the supernatant and the precipitate were analyzed. The precipitate was freeze dried and the dissociation enthalpy and temperature of the AML- complexes were determined using DSC.

THE ROLE OF AMYLOSE-LIPID COMPLEX IN THE ENZYMATIC HYDROLYSIS OF STARCH GRANULES

M. Lauro, T. Suortti and K. Poutanen
VTT Biotechnology and Food Research, P.O. Box 1500, FIN-02044 VTT, Finland .

All non-waxy cereal starches contain lipids roughly in proportion to their amylose content. Barley starch contains about 1% of lipids which are mainly lysophospholipids. Amylose in native barley starch has been suggested as occurring in two forms, as lipid-free amylose (FAM) and lipid-complexed amylose (LAM). LAM has been shown to be more resistant to α-amylase than FAM, but complete digestion of the complex is obtained when a large excess of enzyme or long conversion times are used. Little is known about the recomplexation of lipids and starch in gelatinised and partially hydrolysed starch syrups. The amylose-lipid complex may cause haze, filtering problems and yield losses for example in mashing and in production of starch syrups. In this study, the accessibility of amylose-lipid complex to α-amylase during hydrolysis of granular and partially gelatinised barley starch is examined . The distribution of lipids in the granule residues and leached carbohydrates after different extents of hydrolysis is analysed. The leaching of LAM from α-amylase treated granule residues at different temperatures is also studied.

SIGNIFICANCE OF STARCH GRANULE MINOR COMPONENTS: EFFECT OF SODIUM DODECYL SULFATE (SDS) EXTRACTION

M. Debet and M.J. Gidley
Unilever Research, Colworth Laboratory, Sharnbrook, Bedfordshire, Mk 44 ILQ, England

Differences in many important properties of starch (e.g. gelatinisation and swelling) cannot easily be accounted for by the ratio of its major components, amylose and amylopectin. They therefore depend on either the architecture of the starch granule and (or) the presence (or absence) of minor components, comprising mainly lipids and proteins.

Two of the most crucial features of starch behaviour are (i) the very different swelling properties (rate and extent) exhibited by starches from different botanical origins and (ii) the persistence of granule remnants (ghosts) following complete disruption of polysaccharide order during gelatinisation (Stark et al, 1983; Prentice et al, 1992).

Native starches seem to exhibit two broad types of swelling behaviour: "Rapid and extensive" swelling (root and tuber starches, waxy maize) and "slow and controlled" swelling (cereal, maize and wheat) which tend to correlate with amount of minor components.

Among the main commercial starches, cereal starches (wheat, maize, barley, rice) contain more lipids (0. 6- 1 % w/w) than tuber (potato, 0. 05%), root (tapioca, 0. 1%), legume (less than 0.6%) and waxy mutant cereal starches. This is also true for the protein content : 0.25-0.6% for cereal starches compared with 0.06% for potato and 0.1% for tapioca.

Common commercial starches were washed with a solution of SDS (2% w/v) at 20°C and the effect on starch granule swelling properties was monitored. Rapid swelling starches did not exhibit altered viscosity development profiles, but all slow swelling starches were converted to rapid swelling types following SDS extraction. SDS extracts were characterised by elemental analysis and SDS polyacrylamide electrophoresis. Yields of ghosts from gelatinised maize and wheat starches were not significantly affected by prior SDS extraction, but were more fragile.

References
Prentice, R.D.M., Stark, J.R. and Gidley, M.J. (1992). Granule residues and "ghosts" remaining after heating A-type barley starch granules in water. Car bohydrate Research, 227, 121-130.
Stark, J.R., Aisien, A.0. and Palmer, G.H. (1983). Starch, 35, 73-76.

STUDY OF THE DISSOLUTION OF POTATO GRANULES CONTAINING CATIONIC AMYLOPECTIN STARCH

A. Larsson and S. Wall
Department of Physical Chemistry, Göteborg University. S-412 96 Göteborg, Sweden

In papermaking, retention aids like cationic starch and cationic polyacrylamides play an important role in binding fillers to the cellulose fibers. Dual retention aids like cationic starch and colloidal silicic acid are even more efficient in retaining fillers in the paper sheet. When used in paper applications ordinary cationic potato starch is usually dissolved in jet cookers at 120-130°C for 30 minutes. However, cationic amylopectin starch from potato can be dissolved at lower temperatures, around 70°C.

We have studied the dissolution of 0.20 %wt granular cationic amylopectin starch from potato at 70°, 80° and 97°C at different dissolution times. The granules are added in the solution when the final temperature is reached. After a certain time the solution is cooled with ice to room temperature. Then NaCl is added to the solution in order to contract the amylopectin molecules and make them less shear sensitive. The sodium chloride concentration is 5 mM. The amylopectin solutions have been analyzed using capillary viscometry. Using stopped-flow technique the flocculation between cationic amylopectin and silica particles, with a diameter of 5 nm, has been studied. From this we can estimate the amount of "visible" cationic charges in solution i.e. charges not hidden in the interior of the granules. Light microscopy is used to evaluate whether large granules are more easily dissolved than small ones or vice versa.

MODE OF STARCH TRANSFORMATION DURING COOKING OF DRIED PASTA

B. Conde-Petit, C. Cunin Dalvand and F. Escher
Department of Food Science, Swiss Federal Institute of Technology, Zürich, Switzerland

Although starch presents by far the largest portion of dry matter in wheat pasta, its function as a texturogen starts only by transformation during cooking. The transformation in the starch fraction of dried durum wheat spaghetti during cooking was followed by light microscopy and differential scanning calorimetry (DSC). The water distribution was estimated by determining the water content of thin cuts in the longitudinal direction of spaghetti samples. During cooking dried spaghetti water penetrates towards the center of spaghetti strands. The water content of the outer zone of spaghetti strands rises from 12 to approx. 80 g/100 g within five minutes. In the center of the spaghetti strands the water content rises comparatively slowly and reaches approx. 65 g/100 g in optimally cooked spaghetti (13 min., "al *dente"* point). In optimally cooked pasta the birefringence of starch granules is almost totally lost and no gelatinization endotherm, as measured by DSC, can be detected, indicating a complete melting of the starch crystals. Nevertheless, a continuing change in starch granule structure is observed, which ranges from strongly swollen starch granules in the outer zone to almost no swelling in the center zone of spaghetti strands. Due to the inhomogeneous distribution of water, which acts as a plasticizer, two distinct phenomena related to starch transition occur, i.e. a hydration driven gelatinization process in the "water-rich" outer layer and a heat induced crystal melting in the "water-deficient" center. In any food rich in starch, this dual transition mode has to be taken into account as soon as the moisture distribution is inhomogeneous.

IMPACT OF MULTISTEP ANNEALING ON PASTING CHARACTERISTICS OF WHEAT STARCH

J. Vansteelandt, H. Jacobs and J.A. Delcour
Research Unit Food Chemistry, K.U.Leuven, Kard. Mercierlaan 92,
B-3001 Heverlee, Belgium

The influence of multistep annealing on the pasting behaviour of wheat starch was studied with the Rapid Visco Analyser (RVA). Pasting characteristics as total solubility, swelling power and close packing concentration were determined.

RVA-pasting-profiles were recorded with 6.6 and 9.9% (w/w) starch suspensions. Total solubility, swelling power and close packing concentration were determined after gelatinisation of 1% suspensions of the samples in the RVA.

Multistep annealing definitely has an impact on the RVA-pasting-properties, but the impact depends on the starch concentration used. Results for the annealed samples indicated that multistep annealing caused a higher granule rigidity, a decreased total solubility and swelling power and an increased close packing concentration after each annealing step.

IMPACT OF ANNEALING ON THE SUSCEPTIBILITY OF STARCH TO PANCREATIC ALPHA-AMYLASE

H. Jacobs, R.C. Eerlingen, H. Spaepen and J.A. Delcour
Research Unit Food Chemistry, K.U.Leuven, Kard. Mercierlaan 92,
B-3001 Heverlee, Belgium

Native and annealed wheat and potato starches were subjected to enzymic hydrolysis with pancreatin. Hydrolysis as a function of time showed a decreased enzyme susceptibility of potato starch after annealing. However, for annealed wheat starch, an increased degree of enzymic hydrolysis (after 10h) was observed.

After 2 and 120 h of hydrolysis, starch residues were isolated and characterised by DSC. The gelatinisation behaviour of both native and annealed potato starches was not significantly altered as a function of hydrolysis time. For wheat starch, the gelatinisation behaviour was changed as a result of enzymic hydrolysis. This was more pronounced for the annealed than for the native starch.

POTATO STARCH: DEGREE OF PHOSPHORYLATION RELATED TO DYNAMIC RHEOLOGICAL CHARACTERISTICS.

L. Poulsen* , P. Muhrbeck and J. Adler-Nissen,
Aalborg University, Biotechnology Laboratory, DK-9000 Aalborg, Denmark

Modified starches, including phosphorylated starches have widespread application in the food industry. They are used to control the functional attributes of foods such as thickness and appearance. Potato starch contains varying amounts of phosphorylated amylopectin depending on the type of potato. The influence of phosphorylation on the rheological properties of potato starch pastes and a salad dressing food model system has been investigated with a view to possibly replacing these modified types of starch.

All potato starch samples were ion exchanged with potassium ions to ensure the same ionic environment for all the samples. Some of the chemical parameters of the potato starches, such as the phosphorus content, the size of the starch granules, the amylose percentage and the gelatinisation temperature were determined. The range of phosphorus investigated was 8.1 to 26.3 nmol glucose-6-phosphate/mg starch. Rheological tests were performed using dynamic rheological methods, and the results obtained were expressed as the complex modulus, G^* and the phase angle, δ. These tests were carried out on starch pastes and the model food system. For the starch pastes a measurement of the turbidity of the paste was also performed.

The correlation between the starch paste G^* and starch phosphorus content was found to be positive and linear for the investigated range. Starch paste turbidity could also be correlated to the potato starch phosphorus content. G^* for the salad dressings and the starch phosphorus content were negatively correlated. This correlation was not linear and the starch granule size is expected to effect the G^*.

*Present address: Aalborg University, Biotechnology Laboratory, Sohngaards holmsvej 57, 9000 Aalborg, Denmark.

STRUCTURE AND FUNCTION OF WHEAT STARCH IN REACTION WITH BAKING IMPROVER.

M. Soral-Smietana, J. Fornal, M. Rozad
Division of Food Science, Institute of Animal Reproduction and Food Research
Polish Academy of Sciences, Olsztyn, Poland

Baking is a technological process widely used to influence wheat starch. Baking improver containing wheat flour, glucose, maltodextrins, ascorbic acid and emulsifier (30% C_{16} and 60% C_{18}) was added to wheat flour. Starch interactions on the surface and inside the granule were catalysed by the improver components during dough forming. The intra- interaction was evidenced by the internal amylose ring. Thus, the mechanism may be hypothesized on a limited diffusive-osmotic liberation of amylose from large and medium granules, that results from changes in wheat starch granule structure upon reaction with improver. This mechanism could also be responsible for slower migration of amylose during dough formation and fermentation. However, it could cause the starch structures after the interaction with improver to become more α-amylase- susceptible after baking.

The large granules of starch isolated from improver-made bread 1h after baking were shown to transform easily to a gel phase, and the small granules were entrapped in the gel matrix. On the other hand, the granules of starch from control bread were deformed upon gelatinization and suspended in the matrix of amylose.

By 24h after baking, outer retrogradation-like regions appeared on control bread-starch granules, while an amylose-free area occurred around the granules of starch from improver-made bread. Thus, despite the integrating function, starch affects the redistribution rate of water and its solutes such as amylose fraction, upon the improver ligands.

MECHANICALLY MODIFIED STARCHES AS FAT MIMETICS

C. Niemann and F. Meuser
Technical University Berlin, Department of Food Technology II, D-13353 Berlin

Granular maize starch was partially degraded by enzymic hydrolysis using a bacterial α-amylase to obtain porous starch. In comparison to its native counterpart, the porous starch has different thermal and viscous properties. It is also more susceptible to mechanical damage.

Native and porous maize starch were wet-milled and subsequently spray dried to produce granule fragments. SEM Image Analysis was conducted to reveal the particle size distribution of the fragments. The distributions had a maximum at a particle diameter of 1-2μm. Investigations by DSC and SEC showed a substantial loss of crystallinity and decrease of average molecular weight, respectively, in comparison to the starting materials.

The products are able to form freeze thaw stable gels in cold water on a 20-25% dry matter basis. The gels show little or no syneresis even at high centrifugal forces. Therefore, these modified starches can be used as fat mimetics in various food formulations. Examples for applications in cake toppings, mayonnaises, dressings, etc., will be given .

A STUDY OF WHEAT STARCH DEPOSIT REMOVAL FROM STAINLESS STEEL SURFACES USING CHEMICAL CLEANING AGENTS

R.A. Din and M.R. Bird
School of Chemical Engineering, University of Bath, BA2 7AY, United Kingdom.

Surfaces in contact with food based products require thorough cleaning at regular intervals. The requirements regarding the quality and hygienic properties of the product are increasingly putting more demand on the efficiency of the cleaning process. Insufficient cleaning can lead to product contamination, causing damage or loss of product. In the food industry, cleaning tasks in production processes are fulfilled by fully automatic **CIP** (**C**leaning **I**n **P**lace) systems of high technical standard. The progress in developing cleaning procedures and the criteria for evaluating the effectiveness of cleaning chemicals is of great interest. Considerable time, detergent and energy might be saved if a clear understanding of the principles involved in cleaning starches and a knowledge of the effect of certain variables upon starch removal were determined.

Most studies have concentrated on milk and its products due to the nature of fouling and the reproducible kinetics that have been deduced as a result of this during the cleaning process. However, very little work has been documented on the effective cleaning of starch based soils ocuring in the baking industry. Experimental fouling and cleaning rigs have been designed and constructed to study the mechanisms and evaluate the removal kinetics for cleaning baked wheat starch deposits. The cleaning rig is equipped with a data logging facility to enable continuous monitoring of the parameters affecting cleaning under controlled thermo-hydraulic conditions. Two techniques are employed to measure the kinetics of the cleaning process: one measures   the starch remaining on the surface (gravimetric) and the other measures the starch contained in the effluent stream using TOC (Total Organic Carbon) measurement. Visualisation of the process is possible in a rectangular-sectioned glass flow cell, enabling the use of video camera-aided recording of the cleaning process. Results are presented which show the validity of the experimental protocol, and examine wheat starch removal as a function of cleaning agent concentration, temperature and flow rate. The cleaning agents investigated included sodium hydroxide, nitric acid, *Micro* (a commercial long-chain detergent) and an enzyme-based cleaner. The mechanisms involved in the removal process are discussed, and possible approaches to the modelling of starch removal are suggested.

THE EFFECT OF THE GROWTH ECOZONE UPON THE FUNCTIONAL, STRUCTURAL AND PHYSICOCHEMICAL PROPERTIES OF CASSAVA STARCH : A CASE OF STRUCTURE-FUNCTION RELATIONSHIPS

A. Fernandez, J. Wenham and J.M.V. Blanshard
University of Nottingham, Department of Applied Biochemitry and Food Science, Sutton Bonington Campus, Loughborough, Leicesteishire, LE12 5RD, U.K.
Natural Resources Institute, Central Avenue, Chatham Maritime, Kent ME4 4TB U.K.

Cassava starches of a clearly defined origin showed pronounced differences in their pasting and in the mechanical properties of their gels on aging. Molecular composition and structure as well as granule size seem to be responsible for differences in behavior.

Four cassava cultivars selected from the world collection of cassava germplasm held by the Centro Internacional de Agricultura Tropical (CIAT) were grown in two different ecozones in Colombia. Zone A was characterized by high soil fertility and an average annual daily temperature of 23°C, while zone B was of low soil fertility with an average temperature of 28°C. The cultivars were identified as MPer 196, Mven 25, MBra 881 and CGI-37. All cassava roots were harvested and their starches extracted when the plants were 10 months of age

Starches from plants grown in Zone B compared with those from Zone A had higher aqueous solubilities, higher initial pasting temperatures (Brabender pasting test), developed pastes of lower viscosity, exhibited lower granule disruption and lower gel strength in fresh pastes (Bohlin oscillation test), but the aged pastes (24hr at 2°C) developed a higher gel strength, possessed a smaller granule size, and contained higher levels of amylose (iodine colorimetric test).

The molecular structure was studied using an HPLC size exclusion system on isoamylase debranched starches assisted by a multangle laser photometer. Although the amylose was of slightly higher molecular weight (Mw) from cassava plants grown in zone A than B, this difference was accentuated in the amylose present as an exudate from the starch granules in a 1% aqueous starch paste (Zone A: Mw range $11-13 \times 10^5$; and Zone B: $7.5-8.9 \times 10^5$). It was also observed by size exclusion chromatography that the 1% starch pastes derived from plants grown in zone A had a lower proportion of amylose in solution (7.7-10.7 % of the original starch) than those derived from plants grown in Zone B (12.1-13.4% of the original starch). With the exception of MVen 25, the amylopectins in starches from zone A appeared to have a slightly higher relative population of short A chains and also slightly longer B4 chains than the amylopectins in starches from Zone B.

The higher aqueous solubility of starches from Zone B may be attributed to smaller particle size (greater granule surface area per unit of mass which facilitated the leaching process), to smaller amylose molecules and less branched amylopectin. The higher pasting temperatures and lower pasting viscosity are due to the higher amylose content (as well as to the greater solubility). A high concentration of starch in solution, and the low molecular weight of the solubilised amylose must be responsible for the high gel strength developed in aged pastes prepared with starches from Zone B.

STARCH FILMS WITH VARYING CRYSTALLINITY

A. Rindlav[1], S. Hulleman[2] and P. Gatenholm[1]
[1]Department of Polymer Technology, Chalmers University of Technology, Göteborg, Sweden
[2]ATO-DLO, P O Box 17, 6700 AA Wageningen, The Netherlands

In recent years an increasing interest in biodegradable plastics and packaging has been observed. Many packaging materials consist of cellulosic materials and a plastic film. If the conventional plastic film was replaced by a biodegradable film the whole packaging would become biodegradable. Such a biodegradable film can be made of starch. As a part of a packaging material, the starch film should have certain barrier properties. Starch films are excellent oxygen barriers but are often sensitive to water. A possible way to improve barrier properties is to increase the crystallinity of the material.

We have made starch films from potato starch gels. The gels were allowed to dry under different conditions to form films. X-ray diffraction was used to determine long-range cystallinity of the films. The different films showed a crystallinity ranging from amorphous to 23% crystallinity. Infrared spectroscopy (ATR-FTIR) was also used as means of testing whether differences in short-range crystallinity could be revealed with this technique. Due to different drying conditions and varying crystallinity the films also had varying water contents. An increased water content lead to a decrease in glass transition temperature. The water content and the glass transition temperature both influenced the mechanical properties and the barrier properties of the films. Whilst going from a high to a low air humidity the measured oxygen transmission was significantly decreased due to the reduced plasticizing effect of water. Light microscopy of films dyed with iodine showed different structures for films dried under different conditons.

DETERMINATION OF THE CHAIN LENGTH DISTRIBUTION IN AMYLOPECTIN WITH HPSEC AND HPAEC

H. Fredriksson, K. Koch, R. Andersson and P. Åman
Department of Food Science, Swedish University of Agricultural Sciences, Uppsala, Sweden

The physicochemical properties of starch are to a large extent governed by the amylose and amylopectin ratio and their molecular structure. Knowledge about different types of starch enables the food industry to select suitable raw materials and processing techniques for various purposes. The aim of this study was to chemically characterize five different starches; wheat (cv. Holme), barley (cv. Golf), rye (cv. Motto), pea (cv. Capella) and potato (cv. Desiree). The results will be used as a basis for further research, including processing and nutritional studies, in a joint project between the Department of Food Science in Uppsala, the Department of Food Technology and the Department of Applied Nutrition and Food Chemistry in Lund.

Starch was fractionated by gel permeation and the isolated amylopectin debranched with isoamylase. The chain length distribution of the amylopectin was examined by two different HPLC-systems equipped with a differential refractometer for size-exclusion and a pulsed amperometeric detector for anion exchange chromatography, respectively.

DEBRANCHING OF STARCH USING IMMOBILIZED ENZYMES

G.S. Nilsson, P. Pihlsgård, and L. Gorton
Department of Analytical Chemistry, Lund University, S-221 00 LUND, Sweden

Starch debranching enzymes hydrolyse selectively α-1,6 linkages in branching points but not the linear α-1,4 linkages. By using debranching enzymes followed by size exclusion chromatography the chain length distribution of amylopectin and the non- branched/long-branched amylose content are determined.

A new method, where immobilized debranching enzymes are used instead of using soluble enzymes, has been developed and will be presented here. Pullulanase was covalently immobilized on porous silica beads (Micropil G) and packed in a small reactor inserted in a flow injection system. The major advantages in using immobilized debranching enzymes compared with using soluble enzymes in batch are: the enzymes are reusable, the product is removed from the enzyme momentarily, and often there is an increase in enzyme stability. This leads to a faster, cheaper, and easier analytical method.

Modelling and experimental design were used to evaluate the kinetics of the immobilized enzyme reactor. The system was optimised, and the main effects and factor interactions were evaluated. Different factors such as pH, temperature, and calcium ion concentration were varied at the same time. The immobilized enzymes showed high operational and storage stability without severe loss in activity.

STRUCTURE OF ϕ,β-DEXTRINS DERIVED FROM POTATO AMYLOPECTIN

Q.Zhu and E. Bertoft
Department of Biochemistry and Pharmacy, Åbo Akademi University, BioCity, P. 0. Box 66, SF-20521 Turku (Finland)

Limit dextrins (ϕ,β-LD) of potato amylopectin and intermediate α-dextrins of different size distributions were obtained by the successive action of phosphorylase and beta-amylase.

The composition of unit chains was investigated by gel- permeation chromatography on Superdex 75 of the completely debranched ϕ,β-LD. Long internal chains (d.p. >35) had been hydrolysed by α-amylase into chains with lengths intermediate to the groups of short and long chains. The ratio of A:B chains increased with decreasing size of the α-dextrins. The ϕ,β-LD were also partially debranched with isoamylase into free Bb-chains and Ba-chains that carried A-chains in the form of maltosyl-stubs. The distributions of these chains were similar to those of the completely debranched samples. The ratio of A:Ba increased after alpha-amylolysis, whereas the ratio of Ba:Bb decreased. The characteristic ratio of A- and B-chains obtained from the intermediate products of alpha-amylolysis provide a tool for the investigation of the mode of inter-connection of structural units in amylopectins.

INVESTIGATION OF STARCH FUNCTION USING A NEW METHOD

G. Juodeikiene[1], L. Basinskiene[1], A. Cecnikovaite[1], L. Jakubauskiene[1],
G. Scheining[2] , H. Zenz[2]
[1]Kaunas University of Technology, Radvilenu pl. 19 Kaunas Lithuania and
[2]University of Agriculture, Peter Jordanstrasse 82, Vienna Austria

The need for a rapid, specific, convenient and reliable method for the investigation of starch functionality in cereal and cereal products is widely described. The solution of the problem is concerned with the investigation of starch function and the development of new techniques for its realization. The new method is based on the fact that the cohesive and adhesive forces change during the gelatinisation process. The technical conception of the realization of the method has been developed by periodically measuring the quality of material that sticks to a thin slab. The test has been carried out examinining various model systems: - starch AGENAJEL 20.3 50 (made in Austria, 1993'):- different starch pastes which had various degree sof dextrinization (made in Lithuania, 1994).

The mathematical experiment planning method of type 2^2 was used for investigation of starch AGENAJEL 20.350 . The first factor affecting the experiment is concentration of starch suspension, changed in the range of 4 - 5%; the second factor - time of gelatinisation ranged between 10 and 20 min. The parameters of optimization were rheological jelly, which were evaluated according to the shear forces.

The results of the experiment showed, that rheological parameters during the gelatinisation process correlated well with the amount of sticking. The results of the experiment with starch pastes showed, that the amount of sticking material correlates well with this viscosity. As the dilution of starch pastes increases, the amount of sticking material proportionally decreases. In all tested samples of starch the measured parameter - the amount of sticking material - increases during the gelatinisation process, reaches its maximum when the products reach the gel state and decreases during further heating. It should be noted that the method for the estimation of starch stood out as simple, fast and showed a good reproducibility of the test results .

PROPERTIES OF STARCH FROM WAXY WHEAT

T. Nakamura[1], K. Hayakawa[2], K. Tanaka[2]
[1]Tohoku Nat Agr. Exp. Stn., Akahira 4, Morioka, Iwate 020-01, Japan,
[2]Nisshin Flour Milling Co. Ltd. Research Center, 5-3-1 Tsurugaoka Oi-machi,
Saftama 356, Japan

Starch granules of waxy mutant stained red-brown by iodine while nonwaxy starch stained blue -black or purple, indicating that waxy starch is composed only of amylopectin and lacks amylose which possesses high binding capacity with iodine. Waxy starch shows high swelling power, resistance to gel formation and a reduction in retrogradation. These unique characteristics provide many implications for the food industry. Waxy mutants of durum and common wheat were reported only recently, waxy mutants had earlier been identified in rice and corn, barley and sorghum. At this time we will present the first studies on the properties of waxy wheat starch.

MONITORING OF STARCH GELATINISATION KINETICS BY MEANS OF LIGHT REFRACTION AND RHEOLOGY CHARACTERISTICS MEASUREMENTS

V.N. Golubev and I.N. Zhiganov
Moscow State Institute of Food Industry, Moscow, Russia

Difficulties in direct research method's utilization for studying jelly-like systems led to implementation of a number of indirect methods sensitive enough to internal solution structure alternation - when studying its transition to a jelly-like state.

The gel's internal structure is directly tied to its physical, chemical and optical properties. Indirect methods - optical density and limit voltage displacement measurements - give the best results to study gel properties. These methods are indispensible when investigating molecular chains' alterations in transition from liquid to solid state, deformation, changes in molecular interaction.

The kinetics of potato and corn starch gelatinization have been investigated in concentration range 1-5 %.

Dependencies: $P = f(c_k)$--jelly durability from starch concentration, $p = f(\tau)$ gelatinisation kinetics, starch solutions optical density from starch concentration at $\lambda = 760$ nm, starch's solution light reflection from the light length, starch solution viscosity at different concentrations from temperature and starch solution viscosity from starch concentration were obtained.

The data obtained led to the conclusion that rheological and optical methods give an adequate representation of the starch solution structure formation process characterization.

UK GROWN STARCHES: COMPOSITION AND PROPERTIES

[1]M.F.B. Dale,[2] M.P.C. Cochrane, [1]A.M. Cooper, [2] C.M. Duffus, [1]R.P. Ellis,
[3]A. Lynn, [1]I.M. Morrison, [2]L. Paterson, [2]R.D.M. Prentice, [1]J.S. Swanston,
[1]S.A. Tiller,
[1]Scottish Crop Research Institute, Invergowrie, Dundee DD2 5DA
[2]Crop Science and Technology Department, SAC, West Mains Road, Edinburgh EH9 3JG
[3]Food Science and Technology Department, SAC, Auchincruive, Ayr KA6 5HW

The potential usefulness of any starch to a particular application depends on the range of properties it possesses. In turn, these properties may depend on the structural characteristics of the starch. It is the aim of this project to investigate both the structural and physicochemical properties of a range of starches to see if correlations can be established between starch structure and properties. The investigations are being carried out on cereal and potato starches supplied by industry and on potato, oat, wheat and barley starches from plants grown both in the field and under environmentally controlled conditions.

The chemical and structural parameters to be investigated include starch granule size and structure (Coulter Counter), structure and chain length distribution of amylopectin (methylation analysis and gel permeation chromatography), starch crystallinity (d.s.c.), amylose content (gel permeation chromatography) and phosphorus content (inductively coupled plasma emission spectrometry). Assessment will also be made of starch lipids and starch proteins. Investigations into physical properties such as gelatinisation temperature (hot stage microscopy and d.s.c.), swelling factors, turbidity and paste viscosity will also be undertaken.

The authors wish to acknowledge financial support from the Scottish Office Agriculture, Environment and Fisheries Department.

SOLID-STATE NMR STUDIES ON THE STRUCTURE OF STARCH GRANULES

K.R. Morgan[1], R.H. Fumeaux[1] and N.G. Larsen [2]
[1]Industrial Research Limited, P.O. Box 31-3 10, Lower Hutt, New Zealand
[2]Crop and Food Research Limited, P.O. Box 4704, Christchurch, New Zealand

Recent NMR studies suggest that there are at least three distinct components to a wheat starch granule. (i) highly crystalline regions formed from double-helical starch chains, (ii) solid-like regions formed from amylose-lipid inclusions complexes and (iii) completely amorphous regions, associated with the branching regions of the amylopectin component of starch and possibly the lipid-free amylose. All the components can be observed separately using different solid-state NMR techniques.

IMAGING THE ENZYMATIC DEGRADATION OF STARCH GRANULES BY ATOMIC FORCE MICROSCOPY

A.A. Baker, M.J. Miles, *A.S. Tatham and *P.R. Shewry
H.H. Wills Physics Laboratory, University of Bristol, Bristol, UK and
*Long Ashton Research Station, Long Ashton, Bristol, UK

Atomic Force Microscopy (AFM) is an ideal technique for studying degradation processes. Unlike most other microscopies, AFM combines the abilities to image surfaces directly in three dimensions, to operate in liquid environments and to follow changes to a sample as they occur in "real time". The topographic information obtained can be used to calculate the volume[1] of sample features.

Previous work in this laboratory has shown that AFM can be used to follow enzyme degradation of starch in real time[2]. As a Timmo (wheat) starch granule was attacked by α-amylase, a hole developed from a damaged area of the granule surface, and increased in size as the enzyme attack and imaging continued. The kinetics of this degradation process were then determined by measuring the change in the volume of the hole[3].

In this poster, we will show how the degradation process is affected by the method of imaging (contact mode or tapping mode), and the geometry of the tip scanning the surface. We will discuss the effect of the scanning tip on the process (and therefore the kinetics), and consider the implications of using the AFM in further studies of this type.

[1] Chromosome Classification by AFM Volume Measurement; T.J. McMaster, M. Winfield, A.A. Baker, A. Karp and M.J. Miles; submitted to J. Vac. Sci. Technol. B
[2] Real-Time Imaging of Enzymatic Degradation of Starch Granules by Atomic Force Microscopy; N.H. Thomson, M.J. Miles, S.G. Ring, P.R. Shewry and A.S. Tatham, J.Vac. Sci. Technol. B 12(3), May/Jun 1994.
[3] N.H. Thomson, Ph.D. thesis, University of Bristol, UK.

THE PROPERTIES OF TOBACCO LEAF STARCH

A.H. Schulman, P. Myllärinen[1], T. Tuomi, and K. Poutanen[1]
Institute of Biotechnology, University of Helsinki, Viikinkaari 9, Helsinki, Finland,
[1]VTT Food and Biotechnology, PL 1500, 02044 VTT (Espoo), Finland

Leaf starch, often characterized as "transient starch" by contrast to the "storage starch" of tubers and grains, has received comparatively little attention. We have examined starch quantity, amylose content, and granule morphology of tobacco leaf starch under various growth conditions and light regimes. The granules tend to be small (<5 µm) and the starch low in amylose under all conditions.

THE INFLUENCE OF STARCH-BRANCHING ENZYMES UPON THE STRUCTURE OF LEAF STARCH.

K.L. Tomlinson, J. Lloyd and A.M. Smith
John Innes Centre, Colney Lane, Norwich NR4 7UH, UK

Detailed studies of the two classes of isoforms starch-branching enzyme (SBE) found in the storage organs of higher plants indicate that members of the classes (known as A and B) have distinct properties and could therefore play different roles in the branching of amylopectin *in vivo*. Consistent with this, mutations that specifically eliminate the A class of isoform from maize endosperm (the *ae* mutation) and pea embryo (the *r*-mutation) affect the structure of amylopectin in these organs. However, interpretation of these effects is complicated by concomitant alterations in the timing of expression of SBE activity and in the ratio of SBE activity to that of starch synthase during the development of the embryo. Leaf starch offers a simpler system on which to study the effects of the mutations. The SBE activity of pea leaves is contributed by A and B isoforms, and large amounts of starch are made over a single photoperiod. We have assessed the effects of the *r* mutation on the structure of amylopectin in these leaves, and found that although there is a general increase in average branch length, there is no change in the strongly trimodal pattern of branch length distribution. These data indicate that differences in properties between the A and B isoforms of SBE are not of major importance in determining the polymodal distribution of branch lengths of amylopectin.

PHOSPHORYLATION OF POTATO STARCH

A.M. Bay-Smidt, B. Wischmann, T.H. Nielsen and B. Lindberg Møller
Plant Biochemistry Laboratory, Dept. of Plant Biology, Royal Veterinary- and Agricultural University, 40 Thorvaldsensvej, 1871 Frederiksberg C, Copenhagen, Denmark.

In potato starch phosphate ester groups are bound to carbon atom number 6 and number 3 in the glucosyl residues of the amylopectin fraction in an amount of 16 nmol phosphate /mg starch. The degree of phosphorylation of potato starch can be decreased by omitting or lowering the level of phosphate fertilizer in the growth medium. When no phosphate fertilizer is administered to potato plants grown in a growth chamber in vermiculite, the degree of phosphorylation decreases to 40% of the level in starch from optimally fertilized plants. When tubers from phosphate starved plants are used as seed tubers and re-grown in the absence of phosphate fertilizer, the phosphorylation level is further decreased. Tuber starches from well fertilized plants and from phosphate deprived plants, each separated in 4 grain size classes, serve as a unique tool to investigate the influence of phosphorylation level on the physio/chemical properties without interference from differences in genetic background. Previous results show a 50% increase in degree of phosphorylation in grains < 32μm compared to grains > 71 μm. These data are in agreement with data from literature.

The biochemical pathway for phosphorylation of potato starch is studied by use of intact amyloplasts and stroma derived from the amyloplast preparation. The amyloplast preparation is totally free of contaminating enzymes from the vacuoles and virtually free of cytosolic enzymes. Mithocondrial enzymes are, however, present as a contaminant. Attempts to purify the amyloplast preparation further have so far failed. The preparations have been washed in high salt solutions in concentrations up to 1 M NaCl and in solutions of Triton X-100 in concentrations up to 0.01 % in addition to two washing steps through 2 % Nycodenz. Following treatments, the decrease in activity of amyloplast marker enzymes was higher than the decrease in activity of contaminating enzymes.

By use of intact amyloplasts it has been shown that Glc-6-P is the most likely precursor to be transported across the membrane. In experiments where isolated intact amyloplasts were incorporated with either radiolabelled Glc-6-P or Glc-1-P, the starch synthesis was much more efficient from Glc-6-P than from Glc-1-P. In addition the starch synthesis was dependent on the integrity of the membrane only when Glc-6-P was used as precursor. Starch synthesis from Glc-1-P is therefore most likely catalyzed by starch phosphorylase bound to naked grains.

MANIPULATION OF STARCH BIOSYNTHESIS AND THE STRUCTURE OF STARCH IN TRANSGENIC POTATO PLANTS

G. Abel[1], V. Büttcher[1], E. Duwenig[1], M. Emmermann[2], R. Lorbarth[1], T. Welsh[1], L. Willmitzer [2] and J. Koßmann[2].
[1] lnstitut fur Genbiologische Forschung GmbH Berlin, Ihnestr. 63, D -14195 Berlin; [2]Max-Planck-insthut für molekulare Pflanzenphysiologie, D- 14476 Golm

Starch is synthesized through the "ADP-glucose-pathway", involving ADP-glucose pyrophosphorylase (AGPase), the key regulatory enzyme of starch synthesis, several isoforms of starch synthase and one or two isoforms of branching enzyme. Other enzymes, such as glucano- transferases (D-enzyme), transglucosidases (T-enzyme), hydrolases and phosphorylases may also be important for determining the starch structure and thereby influencing its gelatinization properties.

Transgenic potato plants were generated where starch biosynthesis was manipulated either through repressing the expression of "starch biosynthetic genes" using the antisense RNA approach, or through the ectopic expression of heterologous proteins which are active on α-1,4- glucans.

The antisense repression of the AGPase results in starchless, but sugar-containing, tubers, clearly demonstrating the key role of this enzyme. For the starch synthases the situation is far more complex, since four isoforms (GBSS 1, SS I-III) have been cloned so far. Only for the granule bound starch synthase I (GBSS I) can a clear role in starch synthesis be assigned, as mutants lacking GBSS I or antisense repression result in the production of waxy potato starch. The latter approach was used to investigate the role of the other starch synthase isoforms. The starch quality of the respective transgenic potato tubers is under investigation. The down-regulation of the expression of the branching enzyme surprisingly does not result in the production of high amylose containing starch, however, other structural changes occur, such as a higher degree of phosphorylation. These changes result in a starch with increased viscosities and better gel-forming properties.

The production of pure amylose in plants could be achieved through the expression of the amylosucrase from *Neisseria polysaccharea*, an enzyme converting sucrose into linear α-1,4-glucans. The gene was cloned and the protein was characterised with respect to the product properties. The glucans synthesized are linear with a molecular weight very similar to potato amylose.

EXPRESSION AND AFFINITY PURIFICATION OF BARLEY ADP-GLUCOSE PYROPHOSPHORYLASE IN THE BACULOVIRUS/INSECT CELL SYSTEM.

H. Rognstad, D.N.P. Doan and O-A. Olsen
Plant Molecular Biology Laboratory, Agricultural University of Norway, Norway

ADP-glucose pyrophosphorylase (AGPase) catalyzes the first committed and highly regulated step of starch biosynthesis in all plant tissues. Most plant AGPases, such as one from barley leaf, are activated by 3-phosphoglycerate (3PGA) and inhibited by inorganic phosphate (P_i). However, it has been observed that barley endosperm AGPase is relatively insensitive to these regulators. Insect cells infected with recombinant virus carrying the gene encoding both the large and the small subunit of the barley endosperm, was found to express the barley AGPase, that was enzymatically active. In order to facilitate purification and thus reducing loss, we designed the expressed AGPase so that the enzyme can be purified by a single affinity chromatography step. A six-histidine nucleotide sequence was fused to the 5′ end of the small subunit cDNA. The expressed chimeric AGPase is highly active and shows the same insensitivity to the regulators as the enzyme isolated from barley endosperm. In this poster we will present data on protein purification and on characteristics of the expressed AGPase.

THE BARLEY ENDOSPERM ADP-GLUCOSE PYROPHOSPHORYLASE EXPRESSED IN INSECT CELLS IS RELATIVELY INSENSITIVE TO 3-PGA AND INORGANIC PHOSPHATE REGULATION

D.N.P. Doan, H. Rognstad and O-A. Olsen
Plant Molecular Biology Laboratory, Agricultural University of Norway , Norway.

Full length cDNAs encoding the large and small subunits of the barley endosperm ADP-glucose pyrophosphorylase (AGPase) were co-expressed in insect cell culture. High enzymic activity in the absence of 3-PGA was detected in infected cells (1 .28 units/mg in the Glc-1-P synthesis direction). The expressed enzyme has an apparent molecular mass of approx. 240 kD as determined by gel filtration, and migrates as 2 polypeptide bands of approx. 56 kD and 51 kD on SDS-PAGE corresponding to those of the native barley endopserm AGPase. This suggests that the barley endosperm AGPase is cytosolic and not plastidial. The expressed enzyme was weakly activated by 3-PGA and was not inhibited by inorganic phosphate. Our results confirm that the insensitivity to 3-PGA and inorganic phosphate is an intrinsic property of the barley endosperm AGPase and is not due to proteolysis of the large subunit.

A NEW MUTATION IN PEAS THAT RADICALLY ALTERS AMYLOPECTIN STRUCTURE

J. Lloyd, J. Craig, A.M. Smith, C. Hedley and T. Wang
John Innes Centre, Colney Lane, Norwich, NR4 7UH, UK

A mutation at the rug-5 locus of peas has a dramatic effect on the structure of starch in the embryo. The starch granules appear highly convoluted, and contain up to 50 % amylose. The distribution of branch lengths of the amylopectin is radically altered in a manner which suggests that the structure of this starch may not conform to the "cluster" model. The amylopectin has an abundance of very short branches and also contains a large proportion of chains with dp which may be in excess of 100. Biochemical analysis shows that the main soluble isoform of starch synthase of the embryo is missing in mutant lines and it is likely that the gene encoding this isoform lies at the rug-5 locus. Soluble activity is, however, relatively unaffected because of the appearance of an isoform not detected in wild-type embryos. Isoforms of starch-branching enzyme are unaffected. A model for the structure of amylopectin in the rug-5 mutant is presented, and the implications for our understanding of the roles of starch syntheses and starch- branching enzymes in determining starch structure are discussed.

PHYTOGLYCOGEN PROCESSING: A MANDATORY STEP FOR STARCH BIOSYNTHESIS IN PLANTS

S.G. Ball[1], G. Mouille[1], M-L. Maddelin[1], N. Van den Koornhuyse[1], N. Libessart[2], B. Delrue[1] and A. Decq[1]
[1]Dept of Biological Chemistry, UMR 111 CNRS, Univ des Sciences et Techniques de Lille, 59655 France
[2]Roquette Frères F62136 Lestrem France

Storage starch is usually defined as a mix of 2 distinct fractions: amylopectin and amylose. Amylopectin, the major compound, is composed of intermediate size $\alpha 1,4$-linked glucans (the A and B1 chains) that are clustered together and hooked to longer spacer glucans (the B2,B3 and B4 chains) by $\alpha 1,6$ linkages. Segments of these chains intertwine to form parallel arrays of double helices responsible for the crystallinity of starch. The origin of these arrays resides in the very close packing of the $\alpha 1,6$ at the root of the unit cluster. It has been generally assumed that the clusters were generated by the coordinated action of elongation (starch synthases) and branching enzymes. In order to find out if additional functions were required in order to synthesize such a complex structure, we have isolated starchless mutants from the monocellular alga *Chlamydomonas reinhardtii*. Most of these mutants were affected in genes conditioning the supply of ADP-glucose such as those encoding both large and small subunits of ADP-glucose pyrophosphorylase (*sta1* and *sta6*) or that encoding the plastidic phosphoglucomutase (*sta5*). However 5 distinct mutant strains that were generated by random integration of the arginosuccinate lyase gene (*ARG&*) mapped to a single new *Chlamydomonas* locus (*STA7*) and were characterized by similar pheonotypes. They contained less than 1% of the wild type amount of granular starch and up to 5% of a novel species of soluble glucan. The purified glucans consisted in a high molecular weight $\alpha 1,4$-linked $\alpha 1,6$ branched water-soluble polysaccharide. The number of $\alpha 1,6$ branches estimated by proton NMR amounted to 10%. The iodine polysaccharide interaction displayed a λmax at 490nm similar to that of animal glycogen. We further compared the distribution of chain lengths of the glucans to those of amylopectin, amylose and rabbit glycogen. We found that this polysaccharide was undistinguishable from animal glycogen. Our efforts to understand the enzymological basis of the phenotype has led us to uncover a selective defect in a starch hydrolytic enzyme activity. This enzyme behaves as an α1,6 glucosidase on zymograms. The defect cosegregated in crosses with the mutant phenotype and was found in all *sta7* carrying mutants. The dextrins produced from amylopectin by this hydrolase were analysed by Mass Spectroscopy, TLC and proton NMR. We demonstrate cleavage of the α1,6 linkage in the dextrin product. We thus propose that glycogen is a natural precursor of starch synthesis, that the amorphous lamella of the amylopectin clusters (the root of the cluster) is generated by selective debranching of phytoglycogen and that amylose is synthesized downstream from amylopectin. Such a mechanism implies that in addition to the α1,6 glucosidase other functions may be required to prevent further branching and to trim the crystalline lamella in order to generate mature 9nm clusters. That other genes are indeed involved is suggested by the identification of another *Chlamydomonas* gene (*STA8*) which when defective leads to the simultaneous production of phytoglycogen and high amylose starch.

This work was supported by the Université des Sciences et Technologies de Lille, by the Ministère de l'Education Nationale, by the Centre National de la Recherche Scientifique (Unité Mixte de Recherche du CNRS n° 111. Director André Verbert), by a grant from the Conseil Regional du Nord-Pas-de Calais and the company Limagrain-Ulice and by a special grant from the starch processing company Roquette Frères (Lestrem-France).

Subject Index